Rice Is Nice

108 Quick and Easy
Brown Rice Recipes

New Revised Edition
Introducing the
Planetary Health Food Pyramid

By Wendy Esko

With Forewords by Edward Esko,
Michael Potter, and Gale Jack
and an Afterword by Alex Jack

Amberwaves
Massachusetts • Michigan • California
with Branches from Sea to Shining Sea

To the memory of Aveline Kushi—pioneer, teacher, and friend—who helped introduce organic brown rice to modern society and taught a generation of macrobiotic and natural foods cooks

Rice Is Nice
© 2001 by Wendy Esko

For further information on mail-order sales, wholesale or retail discounts, distribution, translations, and foreign rights, please contact the publisher:

Amberwaves
305 Brooker Hill Road
P.O. Box 487
Becket, MA 01223
U.S.A.

Tel: (413) 623-0012
Fax: (413) 623-6042
Email: info@amberwaves.org

Revised Edition: January 2002
10 9 8 7 6 5 4 3 2 1

ISBN 0–9708913-1-8
Printed in U.S.A.

Contents

Ode to Rice

"O beautiful for spacious skies, for amber waves of grain . . ."
—AMERICA THE BEAUTIFUL

"And God said, Behold I have given you every herb-bearing seed . . . to you it shall be for meat."—BOOK OF GENESIS

"All beings are evolved from food."—BHAGAVAD GITA

"After Siddartha had cleansed himself in the River Neranjara, he took forty-nine mouthfuls of soft rice. After he had the food, he placed the gold cup afloat in the river, making the solemn resolution: 'If I shall become a Buddha today, may this gold cup float upstream.' The gold cup floated up stream for eighty cubits and then sank down."—LIFE OF BUDDHA

"They had best not stir the rice, though it sticks to the pot."
—Miguel de Cervantes, DON QUIXOTE

"When I was about 16 years of age I happened to meet with a book, written by one Tryon, recommending a vegetable diet. I determined to go into it. . . . I made myself acquainted with Tryon's manner of preparing some of his dishes, such as boiling . . . rice . . . I made the greater progress, from that greater clearness of head and quicker apprehension which usually attend temperance in eating and drinking."—BENJAMIN FRANKLIN, AUTOBIOGRAPHY

The Arcadia region of South Carolina "looked like fairyland . . . [it] stands on a sand hill, high for the Country, with [its] Rice fields below; the contrast of which with the lands back of it . . . is scarcely to be conceived."—GEORGE WASHINGTON

"All living things are seen to have spirit, but rice alone is said to have a soul. Djiwa, the soul, lives happily in the fields where rice grows, but has to be enticed to stay with the crop after harvesting."—LYALL WATSON, GIFTS OF UNKNOWN THINGS

"The simplest dishes are the hardest to make. The highest art in cooking is the preparation of a bowl of rice."—MICHIO KUSHI

Foreword

"The most wonderful meals begin with rice."—July, 2001

When *Rice Is Nice* was first published in 1994, the supply of organic brown rice in America seemed secure. Organic rice farming, although small in comparison to chemical farming, was solidly established. The Lundberg family, Ron and Wiladean Houge at Southern Brown Rice, Mike Polit and friends at Polit Farms, and other dedicated growers worked overtime to supply the country with high quality organic brown rice. Organic growers in other parts of the world were working equally hard to provide people in their regions with a reliable source of healthful, natural food. Millions of people—and our planet as a whole—have benefited from their dedication.

The challenge of the past century was to ensure that brown rice and other essential staples were grown free of toxic chemicals, in order to ensure the health of humanity and the environment. Thanks to dedicated growers and pioneer educators and business leaders, that goal was achieved. However, we are now in a new century with new and potentially more serious challenges. Perhaps the most serious of these is the threat of genetically engineered rice, wheat, and other essential foods. How we deal with this challenge could well determine the quality of life experienced by future generations, and quite possibly, the quality of life experienced by all species on the planet. To this end, Amberwaves, a network of concerned citizens, families, farms, and small businesses, has been formed to preserve "spacious skies," "amber waves of grain," and other features of the American dream and our planetary home.

My good friend and colleague, Shizuko Yamamoto, a student of George Ohsawa in Tokyo and for many years a pioneer macrobiotic teacher in America, recently drafted a statement on genetically engineered rice. Shizuko teaches in New York City and around the world and joined Amberwaves as a member of the board of directors. As everyone who knows Shizuko can attest, her views are always clear, intuitive, and direct. Here is what she had to say:

As humankind, we eat grains as our principal food. Recently, we understand that some big corporations want to produce genetically engineered (GE) grain to feed mankind. With GE, mass production, or other scientific techniques, corporations are not reflecting deeply on their benefits to health and nature.

With a clearer view, we can see that everything on this earth belongs to nature, to infinity. Nature's power is beyond our control. At the same time we are a part of nature, as are all animals, plants, and tiny, tiny insects—all are nature's creation. As we know, nature doesn't make anything the same way twice. Everything is unique, even grains of sand on the beach and snowflakes in winter. Each shape is different. Even when identical twins occur, they are not 100% the same. We must understand, acknowledge, and follow the order of the universe. As nature changes, we are going to change, too.

I feel that without GE rice (or other altered grains), we will survive much better. Better not interrupt nature's organic rice pollen with GE pollen, so that we can maintain and preserve the natural order. Our ancestors knew that grains, especially rice, have strong life force and the ability to support many life forms. They observed many creatures living harmoniously every day in the fields. If people are wise enough to appreciate and understand this, please join us and give your support to Amberwaves in whatever way possible.

Last summer, I made several trips to Northern California. While there I spent time in the rice fields of the Sacramento Valley. I also met with organic rice farmers. I visited the rice at different stages in its development. In the Sacramento Valley there are rice fields as far as the eye can see. Rice farming began in the 19th century as Chinese immigrants began cultivating their staple grain. Wheat was the principal cereal crop in the area, but after several droughts, the wheat crop collapsed and farmers shifted to rice. Rice depends on a steady supply of water, and the successful channeling of fresh mountain water from the High Sierras through a vast network of canals made rice farming flourish. The combination of bright sunshine, fertile soil, and fresh mountain water is ideal for growing rice.

My experience in California was similar to what I experienced years before in rice fields in Japan. As are all the cereal grains, rice is a flexible, elegant, intelligent plant. Being in the fields gives one a sense of the profound harmony of Great Nature, of what the Japanese call *Daishizen*. One senses the movement of life from the infinite to the finite world of nature, and the beauty, harmony, and interconnectedness of it all. One becomes aware of an underlying order to creation and one's place within it. In me, that awareness brings forth a deep sense of awe and gratitude for the miracle of life.

In Sanskrit there is a word *Chetana* that means "innate intelligence in food." Nowhere is this innate intelligence more apparent than within brown rice and other cereal grains. Whole organic brown rice possesses a high degree of innate intelligence—for example, an ideal nutritional value—while refined white rice is a food stripped of its innate intelligence. Although genetically intact, chemically produced rice is a food in which the innate intelligence of the species has been weakened by chemicals. Modern genetically engineered rice is an unnatural "food" in which the innate intelligence of the species has been violated, permanently distorted, and quite possibly destroyed.

All natural species possess innate intelligence, while modern genetically engineered "species" do not. Instead, they possess a form of artificially induced intelligence, a derangement resulting from the alteration or rearrangement of their genetic material.

What we face today is a struggle between wisdom and cleverness. Wisdom sees the innate intelligence in nature and the value of cooperating with it. Cleverness, on the other hand, is blind to the innate intelligence in nature. It foolishly believes human beings have the power to override nature without serious consequences to their own health and well being.

Wendy and I studied cooking many years ago with Aveline Kushi. Aveline, who passed away this summer at the age of 78, is probably best known for her introduction of the art of cooking for health and spirituality. Her cooking was artful and delicate, yet healtfhul and strengthening. She had an intuitive understanding of food and nature, and an intuitive mastery of the art of cooking. Aveline referred to her art as the art of macrobiotic cooking, from the Greek *macro*, meaning "large" or "great," and *bios*, meaning "life." As Aveline explained it, to live according to macrobiotic principles meant to live a large or great life.

When it came to food and cooking, Aveline was a purist in the true sense of the word. She insisted on only the finest natural and organic ingredients, pure spring water, cooking utensils made from natural materials, and a natural gas flame. She worked with these raw materials to create true masterpieces in the kitchen. To her, cooking was the highest form of art.

Although all of her dishes were healthful and delicious, perhaps her greatest achievement was her brown rice. Aveline's brown rice was close to perfect. Her rice always had a wonderful glutinous consistency and an incredibly rich sweet taste. With every chew, it became more delicious and satisfying. Even today, when I cook brown rice, I strive to achieve the perfection I experienced at her table. Aveline's brown rice is the standard upon which I evaluable the success or failure of my own cooking.

From my own experience, I know that cooking rice on a daily basis creates order and harmony in meals. On the many days when I begin cooking with no specific plan, I remind myself, "Start the rice first." After measuring the rice, washing it, adding water and salt, and placing it on the burner, the outlines of the meal come into focus. One by one, a variety of complementary vegetable and other side dishes takes form, almost without effort.

This classic book, *Rice Is Nice*, is designed to help you experience *Chetana*, the innate intelligence in food and nature. It introduces a wide range of rice-based dishes that can serve as the foundation for a healthful, ecological, and sustainable cuisine. It has been updated to reflect ongoing refinements in the all-important art of cooking, for example, less emphasis on pressure cooking and more on boiling. Healthful cooking is not a rigid set of rules but a constantly evolving process based on everyone's unique perspective and experience.

It is my hope that *Rice Is Nice* will help us all awaken our innate intelligence and realize the path to genuine health and happiness lies in cooperation with nature.

Edward Esko
Becket, Massachusetts
August 2001

Edward Esko, Wendy's husband, is an international lecturer, dietary counselor, and author of Healing Planet Earth, Contemporary Macrobiotics, *and other books. He is Vice President of Amberwaves.*

Foreword

"By his activity in this fight he brought down on his head the lasting hatred of all the food adulterators in America. Until this day no name is more abhorrent to them than that of Theodore Roosevelt. And that organization of food adulterators is one of the most powerful political influences the country has ever had to deal with. It has openly defied the national and state governments for a quarter of a century. The enormous profits it realizes from the processes of imitation and adulteration permit it to have a mighty barrel to spend in influencing legislation . . ."
—"THEODORE ROOSEVELT AND PURE FOODS," *NATIONAL FOOD MAGAZINE*, SEPTEMBER 1912

The more things change, the more they stay the same.

Today we face a more permanent adulteration of our food, the pollution of our food gene pool by giant corporations which own the DNA and are only interested in profit. This affront to organic agriculture and enormous burden on companies like Eden Foods to acquire non-genetically polluted organic food, is the most serious threat to the freedom of humanity from corporate exploitation ever. It strikes me as ironic that food would be genetically engineered to allow the killing of all life in the soil with toxic herbicides, when the inherent nature of anything considered food is to nurture life.

These same corporations are, today, actively working to acquire ownership of all the fresh water in India and Mexico, with plans for owning all fresh water.

Nature created perfect food. It is arrogant to think that a scientist can make food better by manipulating DNA. It is far more likely that this tampering will limit the expression of the genetic potentialities of life. Recklessly releasing man-made, untested organisms into our environment may lead to surprising, devastating results, and has already caused enormous economic damage.

For a frightening look at the results of man introducing designed-for-profit DNA into the environment, one need only look at the deplorable condition of the salmon species and the rapid disappearance of all wild salmon in Scandinavia, the U.K., Nova Scotia, Washington State, and British Columbia.

Genetic science is not itself bad. It has potential for creating new silver bullet medicines of value. Tampering with the DNA in our food supply is another thing, and the overwhelming majority of the people on Earth do not want it done.

I have stood on huge tracts of dead soil, all life choked out of it by agricultural chemicals made by these same corporations. For Nature to repair this requires thousands of years; but, we are "the salt of the Earth," and the magic of man is that we can repair this damage using organic techniques in three to five years. What power we have to nurture life, or kill, if we do not act wisely and prevent further processes of imitation and adulteration of food.

Michael Potter
Clinton, Michigan

Michael Potter, founder and president of Eden Foods, is a pioneer in the natural, organic foods movement. He is on the Board of Directors of Amberwaves. Established in Ann Arbor, Michigan in 1968, Eden is a rare independent pioneer natural food company. Eden has created a dedicated network of family farms and suppliers, nurturing more than 30,000 acres of family owned, organic farm land and offers over 182 organic and traditional foods.

The Eden brand means: no irradiation, no preservatives, no chemical additives, no food colorings, no refined sugars, no genetically engineered ingredients—the safest, most nutritious, certified organically grown food that can be offered. www.edenfoods.com

Foreword

On Thanksgiving Day, in 1949, two events occurred three thousand miles apart that would change America's destiny. Michio Kushi set foot on American soil for the first time in San Francisco, and a baby girl was born in upstate New York. Michio went on to become the leader of the macrobiotic, natural foods, and holistic health movements, while Wendy Esko became the leading macrobiotic cook born in North America.

I first studied cooking with Wendy in 1980. I discovered macrobiotics at a seminar Michio and Aveline, his wife, gave at a church in Dallas. For several years I struggled to cook for my young son and myself. It soon became clear that I needed experienced guidance, and I enrolled in the Kushi Institute, then headquartered in Brookline, Massachusetts. Wendy was my first cooking teacher and later, in 1985, when I finished the last level of the Leadership Training Program in Becket, MA (current home of the Institute), she taught every cooking class in my course. Her food was uniformly delicious, attractive, and energizing. I remember mostly the noodle and deep-fried vegetable dishes, the seaweed casseroles, and the scrumptious desserts, but the foundation of her cooking was always brown rice.

The style of cooking that I learned with Wendy enabled me to keep my—at that time—fragile health and cope with the pressures of being a single mother. Within several years, I had remarried and with my husband, Alex, who also taught at the K.I., I spent a summer living at the Esko house in Becket. I often cooked with Wendy, pounding mochi, rolling sushi, and making other dishes together, and gradually developed my skills and confidence.

Wendy had a tribe of five children at the time (she currently has eight), so her composure and concentration in the kitchen struck me as all the more remarkable. Like Wendy—who had some Iroquois ancestry—I had some Cherokee influence on my father's side, and I think this background brought us closer together, as well as helped root us in the American soil. Wendy has been a pioneer, not only in introducing macrobiotic cuisine to America, but also in shaping, developing, and adapting principles of balance and harmony to this continent's unique climate, environment, and energy flow.

Wendy has written many wonderful macrobiotic cookbooks over the years. In *Rice Is Nice*, her first cookbook devoted entirely to brown rice, Wendy shares over 100 of her best recipes with us. Here is everything from basic pressure-cooked brown rice to rice and bean casseroles, rice croquettes, rice balls, rice paella, rice porridge, and mochi waffles.

In Greek mythology, Athena, the goddess of wisdom, sprang fully formed from the Zeus' forehead. In our secular age, we no longer believe in myths, fairy tales, and the arrival of mysterious culture-bearers. But the appearance of Michio and Wendy in America—on Thanksgiving, the archetypal macrobiotic holiday—can only make one wonder whether our beautiful country is still governed by a wise Providence. I hope that you enjoy this inspired book. I know your friends and family certainly will.

Gale Jack
Becket, Massachusetts

Gale Jack is a macrobiotic cooking teacher and author. Her books in-clude Promenade Home, Amber Waves of Grain, *and* Women's Health Guide. *She is treasurer of Amberwaves.*

Preface

I was raised in upstate New York, where the rich soil and rolling hills are perfect for growing grains, beans, and vegetables. We lived next to a farm, and it was always a pleasure to sit on our front lawn and observe the fields of grain across the way. Our neighbors grew wheat, oats, and barley. I would watch as they plowed, planted, and eventually harvested the grain. After planting, I would watch the tiny sprouts grow into mature plants. At first the fields were light green, then slowly turned golden as the grain ripened and became ready for harvest.

I loved watching the wind blow across the fields. Grain plants are so flexible and yielding. They would bow down and spring back, reminding me of waves on the Great Lakes. It was so peaceful and calming. Often my siblings and I would walk or play hide and seek in the fields, knowing that soon our beautiful playing fields would be cut and sent off to provide food for people. The grain in the fields brought forth images of steaming bowls of oatmeal on cold winter mornings, the aroma of freshly baked breads and pastries, and the rich creamy texture of thick barley-vegetable soups.

As a child I loved to eat grains. I remember asking my mother to let me eat some of my baby brother's rice cereal. I looked forward to her barley soups and grain-stuffed cabbage dishes. On Fridays, it was rare for me to be at home for dinner. Our neighbors, who were from the South, served white rice every Friday. They knew I loved rice and often invited me to join them for dinner.

I tasted my first bowl of brown rice twenty-four years ago. I was impressed by how different it was from white rice, so chewy and sweet. One of the first things I noticed was that brown rice did not leave an aftertaste in the mouth. Rather, it seemed to make the mouth feel cleaner. Unlike white rice, which I ate without feeling satisfied, brown rice was satisfying and complete. I was so impressed by the taste of brown rice that after attending a macrobiotic cooking class in 1971, I readily understood the advantages of a grain-based diet.

In 1978, my husband, children, and I journeyed to Japan. We delighted in watching rice being planted in small paddies around Kyo-to. I watched the wind blow across the rice fields, creating the same rolling waves I had seen as a child. We would observe the fields as if in meditation. The effect was very peaceful and calming. After seeing the rice, I felt grateful to nature and to the people who worked hard to provide our family with the grains on our table. I also felt thankful for the good health these foods had given us.

Along with cooking for my family over the years, I have cooked for thousands of students and friends. I have also given many classes on macrobiotic cooking. With every pot of rice I have cooked, I experienced the same calm, peaceful energy I felt when seeing brown rice and other whole grains in the fields.

Michio Kushi has said that there are at least a thousand ways to prepare and serve brown rice. In this small book, my first for Amberwaves, I present a sample of these possibilities, beginning with the most basic methods. The basic methods of cooking brown rice are like notes in a musical scale. Hundreds and even thousands of variations can be created from them. I hope you use these recipes to create your own delicious and healthful dishes.

I would like to thank everyone who inspired this book. I thank my teachers, Michio and Aveline Kushi, who pioneered the introduction of brown rice and other whole natural foods in North America, Europe, and throughout the world. I thank my husband, Edward, for helping with the manuscript, and Gale and Alex Jack for their overall guidance. I thank Michael Potter, president of Eden Foods, for his encouragement and support. I thank my students, children, and friends for their love, support, and inspiration.

Wendy Esko
Clinton, Michigan

Introduction

"Among cereal grains, brown rice is the most balanced. Its size, shape, color, texture, and proportion of carbohydrate, fat, protein, and minerals fall in the middle of the spectrum of the seven principal grains. Rice is biologically the most integrated grain—our evolutionary counterpart in the plant world."—
Aveline Kushi

Over the last generation, brown rice has become a staple in many households across North America and Europe. The health benefits of eating brown rice are now recognized by doctors and nutritionists around the world. This ancient grain is the most widely used food crop on the planet. For nearly 3 billion people in Asia, rice provides 25 to 80 percent of their daily calories. As the 21st century began, about 600 million metric tons of rice were harvested annually around the world, nearly all for human consumption. Although the wheat and corn crops were slightly larger, these grains are often used as livestock feed. About 90 percent of the world's rice crop is grown and eaten in Asia, the majority in China and India.

Much of the rice harvested today is processed and sold as white rice. Brown rice is a nutritionally complete whole food. White rice is not. All rice is brown until the food processors use a process called "pearling" to remove the mineral- and vitamin-rich shell or skin. White rice is mainly starch and does not contain the natural vitamins and minerals found in brown rice.

Cultivated rice, or *Oryza sativa*, is grown in standing water, or paddies, and on dry land. Low-land rice fed by rainfall comprises about 25 percent of the world's rice crop, while paddy rice comprises about 55 percent. (The word *paddy* is a Malaysian term for rice growing in deep water. It also refers to "rough," or unhulled rice.) Rice is also planted on forest slopes that have been cut and burned. This type of rice is known as "upland" rice. There is a substantial difference in flavor and energy between paddy and dry land rice.

Rice is also divided into two primary strains, based on the climate of origin. More hardy rice, or *japonica*, originated in the temperate zones. Sometimes referred to as "sticky" rice, it has short wide grains that stick together during cooking. Japonica rices have a strong sweet flavor. (California short grain brown rice is an example of japonica.) They are relatively resistant to temperature changes while growing, but are sensitive to the amount and duration of sunlight. More delicate rices, known as *indica*, originated in India and are grown in the warmer, more tropical regions. Indica rices are sensitive to temperature changes, but not to light. Basmati rice is an example of indica. Several of the more common varieties of rice are described below, and are categorized into staple rices, or those recommended for regular use, and exotic, or specialty rices suitable for special occasions.

Staple Rices

Short Grain Brown Rice Short grain brown rice is the most suitable for regular use in a temperate or four-season climate. It is the smallest, most rounded, and hardiest of the main varieties of rice. Short grain brown rice contains a high amount of gluten, which is the protein factor in the rice, along with a higher concentration of minerals. It has a naturally sweet taste and a balanced proportion of minerals, proteins, carbohydrates, and fats. The recipes in this book call mostly for the use of short grain, organic brown rice. Other varieties can be substituted according to your needs and desires.

Sweet Brown Rice Sweet brown rice is higher in gluten than the other varieties of rice. It is also slightly sweeter in taste. Sweet brown rice is often used in making mochi, a sweet taffy made by pounding the grains into a thick, sticky paste. Sweet brown rice is also used to make ohagi, a soft, sweet rice patty coated with such things as roasted walnuts, sesame seeds, sweet azuki beans, or cooked and mashed chestnuts. Sweet brown rice is used in making dumplings for soup, in making cookies, crackers, and a sweet drink called amazake.

Medium Grain Brown Rice Medium grain brown rice is slightly softer in texture than short grain rice. It is lighter and not as sweet as short grain rice. Long grain brown rice is fluffy and light when cooked. It is more suitable for use in semitropical climates or in the temperate zones when the weather is hot. It makes excellent fried rice because of its light, fluffy quality.

Long Grain Brown Rice This is the lightest of the major varieties of brown rice. It has a chewier texture than long grain white rice, and contains fiber and vitamins. Long grain rice is more suitable for use in warm weather.

Special Occasion Rices

Wild Rice Wild rice, an uncultivated aquatic grass used by native Americans, is not a member of the same species as regular rice but shares similar qualities. It is used most often on special occasions or holidays. The plant is very tall and robust. Usually the grains are sun-dried or parched by heating in a metal drum. It can be used in soups, stews, stuffings, puddings, or fried rice. Much of what is sold as "wild" rice is actually cultivated and harvested and processed by machine. Hand-harvested lake- and river-grown wild rice is more flavorful than the cultivated variety.

White Rice White rice has the outer mineral- and vitamin-rich bran removed. Refined grains and their products are generally not suitable as staple foods for optimal health, but can be used for variety as occasional supplements to whole grains. Look for white rice that is organically grown and does not contain talc. Organic white rice can be used in making sushi, fried rice, soups, and stews. Like brown rice, white rice comes in short, medium, and long grain varieties. It also comes in "sticky" varieties that clump together when boiled or steamed, such as the "sticky" long grain rice used in Chinese cooking and the Japanese "sticky" short grain rice used in making sushi. Long and medium grain white rices tend toward blandness, and are often used to absorb the flavors of other ingredients in stir-fries, soups, pilafs, stuffings, and other dishes.

Brown Basmati Rice This long grain aromatic rice is grown in India and the United States. The American variety--known as texmati--is grown in Texas. Brown basmati is more nutritious than white basmati, and has a crunchier texture and slightly nutty flavor. It is more suitable for use in semitropical or tropical climates, or for variety in a temperate climate when the weather is hot.

White Basmati Rice Originally from India and Pakistan, white basmati rice has a delicate nutty aroma. When boiled, it expands only in length, not in width, and the grains remain separate. The hybrid basmati grown in Texas is less pungent and aromatic that the native Indian variety. Other Indian rices include sambal, a short grain non-aromatic variety from Sri Lanka and southern India, and gobin ghog, a short grain, aromatic rice grown in North India. Gobin bhog is similar to basmati but with a stickier texture.

Wehani Rice This dark brown long grain rice is grown by Lundberg Farms in northern California. It has a nutty aroma and can be used in stuffings, salads, and pilafs.

Black Japonica This hybrid rice was developed by Lundberg Farms by combining a dark brown medium grain rice with a black

Japanese short grain variety. It adds an exotic touch to stuffings, stir-fries, and side dishes.

Red Rice This species of rice, originally from Asia, and now grown in the United States, has a red outer coat. It is suitable for warmer climates or during hot weather, and adds a firm, chewy texture to salads and casseroles. A variety of red rice, known as Sri Lankan Red, has a mild flavor and cooks quickly and is often used with seafood dishes. A short grain variety, known as Lundberg Christmas Rice, has a distinctive nutty, roasted flavor. The grains don't stick together when cooked. It makes nice stuffings, puddings, and desserts.

Thai Rices A variety of rices from Thailand are now available. These include jasmine, a long grain rice with an aroma like that of basmati, Thai sticky, a variety that sticks together when cooked, and sticky black rice. When cooked, this glutinous variety has a dark purple color. People in Thailand use it mostly for desserts.

Italian Rices Rice has been used for centuries in Italian cuisine. The more common varieties include arborio, a translucent rice with a white dot at the center, a highly refined variety known as carnaroli, and maratello, a medium-long partially refined rice that cooks quickly. These rices can be used in risotto and Italian-style soups.

Valencia This variety was introduced to Spain by the Arabs. It is related to the Italian rices and is traditionally used in paella.

Louisiana Rices Speciality rices from Louisiana include wild pecan, a product of the Cajun rice growing area with taste like that of pecans, and della, a hybridized cross between long grain varieties and aromatic basmati.

The recipes that follow introduce the many uses of brown rice, especially the organic short grain brown rice which is most suitable for use in temperate climates. Because of its naturally sweet flavor and chewy texture, rice is the one grain that can be eaten daily in its whole form. And, since it is such a versatile food, it can serve as the basis for an endlessly varied healthful cuisine. As you start cooking with brown rice, I hope you will begin to experience a sense of joy and happiness like that expressed in the following Chinese poem (from the eighth century BC):

> We pound the grain, we bale it out.
> We sift, we tread,
> We wash it--soak, soak;
> We boil it all steamy...
> As soon as the smell rises
> God on high is very pleased:
> "What smell is this, so strong and good?"

1
Basic Brown Rice

"Always reflect on the quality of your brown rice and other staple foods, and always seek to improve your cooking. Approach your cooking with a beginner's mind."—Michio Kushi

The quality of water used in cooking has a tremendous effect on the flavor and energy of the dishes you prepare. The best quality water is good, clean well or spring water which is free of chemical pollutants or additives. If well water is not available, good quality spring water can usually be purchased in natural food stores or from spring water companies. Distilled water is lifeless and unnatural, and is not recommended for cooking or drinking.

Usually, a tiny, two-finger pinch of white sea salt is recommended for each pot of grains. White sea salt is rich in trace minerals. In some cases, a small piece of kombu sea vegetable, which is also rich in minerals, can be used instead. The kombu is first soaked for 3 to 5 minutes and then diced before you place it in the cooking pot. A piece of kombu the size of a postage stamp can be used on occasion. Sea salt makes whole grain dishes easier to digest.

Prior to washing your grains, beans, seeds, or nuts, first sort them to remove any small stones, clumps of soil, or badly damaged pieces. Place the grains, beans, seeds, or nuts in a bowl, place the bowl in the sink, and fill it with cold water beyond the level of the food. Rinse by stirring gently with your fingers, and pour the water off. Repeat the process again, and then transfer the food, a handful at a time, to a strainer. Rinse quickly under cold water. Your grains, beans, seeds, or nuts are now ready to be cooked or roasted.

To soak whole grains, after washing them in cold water, place them in a bowl or pressure cooker. Add the required amount of water, as instructed in the recipe, and soak, without salt, for 6 to 8 hours. Whole grains can be cooked in the water you use to soak them.

Boiling and pressure cooking are energizing ways to prepare short grain brown rice. They approximate traditional methods in which short grain brown rice was cooked in a heavy pot with a lid made of thick wood. They help the grains retain energy and nutrients. Pressure cookers are safe and easy to use. Boiling and pressure cooking are the two most common methods for cooking rice.

Boiled Brown Rice

Boiling brown rice creates a light, fluffy dish. Boiled rice can be prepared when you desire light, upward energy in your cooking. Boiling is often appropriate for daily cooking.

> **3 cups brown rice, washed**
> **6 cups spring or well water**
> **small pinch of sea salt**

Place the rice and sea salt in a heavy pot. Cover with a heavy lid. Bring to a boil on a high flame. Reduce the flame to medium-low and simmer for 50 to 60 minutes. Remove with a wooden rice paddle and place in a serving dish.

Basic Pressure-Cooked Brown Rice

This is the most basic method for pressure cooking brown rice. Because the rice is brought up to pressure right away, it has a stronger, more condensed quality of energy. This method is thus more appropriate when stronger energy is needed from your foods. On the other hand, boiling is more appropriate when lighter, more relaxing energy is required.

> **3 cups organic short grain brown rice, washed**
> **4 1/2 cups spring or well water**
> **small pinch of sea salt**

Place the washed rice, water, and small pinch of sea salt in the pressure cooker. Fasten the lid on the cooker and place the cooker over a high flame. When the pressure comes up, place a flame deflector under the cooker and lower the flame. Cook for 45 to 50 minutes.

When the rice is done, remove the cooker from the stove and let the pressure come down. Remove the lid and use a wooden rice paddle to scoop the cooked grain into a serving bowl.

When serving, use the rice paddle to scoop individual portions of brown rice onto each person's plate, or the serving bowl can be passed from person to person and each person can help themselves. Leftover rice can be stored overnight in the pantry, refrigerator, or on a kitchen counter.

Quick-Soaked Pressure-Cooked Brown Rice

This method softens the rice before cooking. The rice cooks more thoroughly and has a naturally sweet taste.

> 3 cups organic short grain brown rice, washed
> 4 1/2 cups spring or well water
> small pinch of sea salt

Place the washed rice in the pressure cooker and add water and sea salt. Place the uncovered cooker over a low flame until the water just starts to boil. This will take 10 to 15 minutes, depending on the amount of rice in the cooker. If you are using sea salt, add now, and place the lid on the cooker. Turn the flame to high, and bring up to pressure. Reduce the flame to medium-low, and place a flame deflector under the cooker. Cook for 45 to 50 minutes.

When the rice is done, take the cooker off the burner. Let the rice sit for about 5 minutes. Place a chopstick under the gauge on the lid of the pressure cooker, thus releasing pressure more rapidly. Remove the lid and use a wooden rice paddle to scoop the rice into a serving bowl.

Pre-Soaked Pressure-Cooked Brown Rice

In this method, the rice is soaked from 6 to 8 hours or overnight. Water causes the rice to become expanded and fluffy. This more expansive quality of brown rice can help balance a dry hot climate or hot summer weather.

> 3 cups organic short grain brown rice, washed
> 4 1/2 cups spring or well water
> small pinch of sea salt

Place the washed brown rice in a bowl and cover with the

amount of water mentioned above. Cover the bowl to prevent dust from entering and set it aside to soak for 6 to 8 hours or overnight.

Place the soaked rice, water used for soaking the rice, and sea salt in a pressure cooker. Fasten the lid on the cooker and set over a high flame. When the pressure comes up, reduce the flame to medium-low, and place a flame deflector under the cooker. Cook for 45 to 50 minutes.

When the rice is done, remove the cooker from the stove and let it sit for 5 minutes. Let the pressure come down naturally or place a chopstick under the pressure gauge. Remove the lid, and use a wooden rice paddle to scoop the cooked rice into a serving bowl.

Pressure-Cooked Roasted Brown Rice

Roasting the grain before you cook it produces a drier, fluffier dish of brown rice. Roasting concentrates energy in the grains, and can be used to help balance humid weather and on special occasions.

3 cups organic short grain brown rice, washed
4 1/2 cups spring or well water
small pinch of sea salt

Heat a stainless steel skillet over a high flame. When hot, place the washed and drained rice in the skillet. Use a wooden rice paddle or spoon to dry roast the rice, moving it constantly back and forth until most of the water has evaporated. Reduce the flame to medium, and continue roasting for several minutes until the rice releases a nutty fragrance and turns slightly golden. (Be careful not to scorch the grains.)

When the rice has been thoroughly roasted, remove from the skillet and place in a pressure cooker. Add water and sea salt, and fasten the lid on the cooker. Turn the flame to high, and let the pressure come up. Reduce the flame to medium-low and place a flame deflector under the cooker. Cook for 45 to 50 minutes on a low flame, then remove the cooker from the stove and let the rice sit for 5 minutes. Let the pressure come down and remove the lid. Serve as described above.

Ohsawa Pot Pressure-Cooked Brown Rice

The Ohsawa pot is an earthenware pot with an earthenware lid. It is named after George and Lima Ohsawa, who were the founders of the contemporary macrobiotic movement. The washed rice, sea salt, and usual amount of water are placed in the pot. The lid is fastened on the pot, and the pot itself is placed inside a pressure cooker which has about an inch of water in it. The lid is then fastened on the pressure cooker, and the cooker is placed over a high flame and brought up to pressure. When the pressure comes up, the flame is reduced to low and the rice is allowed to cook for 45 to 50 minutes. This method concentrates energy in the brown rice.

2
Brown Rice
with Other Grains

"Short-grain brown rice is ideally balanced, particularly for people living in temperate climates. Use a rich variety of staples—rice, buckwheat, wheat, millet, barley, rye, oats, and corn—selecting what grows locally and has been traditionally enjoyed in your part of the world. The grains you eat should be organically grown, free of chemical fertilizer and poisonous spray."—Lima Ohsawa

The usual proportion of rice to other grains in combination dishes is 3/4 to 2/3 brown rice to 1/4 to 1/3 of the other grain. For optimal variety and balance, it is recommended that you combine your rice with other grains on a regular basis. When combining other whole grains with brown rice, it is sometimes necessary to soak, roast, or boil them first.

By combining brown rice with other grains you can create a variety of energies in your primary grain dish, while also providing a variety of different flavors and textures. Some grains, such as whole corn or hato mugi, have a slightly bitter flavor. Other grains, such as fresh sweet corn, sweet brown rice, millet, and whole oats, have a mild, subtly sweet flavor, while others, such as whole barley, wheat, and rye, have a chewier texture. Other grains also add protein, minerals, and other nutrients to your brown rice dishes.

Although the recipes in this chapter call for the use of a pressure cooker, they can just as easily be prepared by boiling in a pot with a tight-fitting lid. Boiling is especially recommended for those seeking lighter, more relaxing energy in their cooking.

Brown Rice with Barley

Barley has a light, upward quality of energy. Adding it to brown rice makes the dish fluffier and less glutinous. Barley can be cooked with brown rice on a regular basis.

    2 cups organic brown rice, washed
    1 cup whole barley, washed and soaked for 6 to 8 hours
    4 1/2 cups water, including water used to soak barley
    small pinch of sea salt

Place the brown rice, barley, and water in a pressure cooker. Place the uncovered cooker over a low flame until the water starts to boil. Add sea salt at this time. Cover the cooker, turn the flame to high, and bring up to pressure. Reduce the flame to medium-low and place a flame deflector under the cooker. Cook for 45 to 50 minutes. Remove from the flame, and allow the pressure to come down. Remove the cover and allow to sit for 4 to 5 minutes. Remove the rice and barley from the cooker and place in a serving bowl.

Brown Rice with Pearl Barley

Pearl barley, or *hato mugi*, is valued in Oriental countries for its power to neutralize the harmful effects of animal food. It adds a wonderfully light quality to your brown rice dishes.

    2 cups organic brown rice, washed
    1 cup organic hato mugi, washed
    4 1/2 cups water
    small pinch of sea salt

Place the brown rice, hato mugi, and water in a pressure cooker. Place the uncovered cooker over a low flame until the water just begins to boil. Add the sea salt, cover, and turn the flame up high. Reduce the flame to medium-low and place a flame deflector under the cooker. Cook for 45 to 50 minutes. Remove the cooker from the flame and let the pressure come down. Remove the cover and let the grains sit for 4 to 5 minutes before placing in a serving bowl.

Brown Rice with Wheat Berries

2 cups organic brown rice, washed
1 cup organic soft spring or pastry wheat berries,
    washed
4 1/2 cups water
small pinch of sea salt

Heat a stainless steel skillet over a high flame. Add the washed and drained wheat berries. With a wooden spoon or bamboo rice paddle, stir constantly to ensure even roasting and prevent burning. When the wheat berries are done they release a sweet, nutty fragrance, turn slightly golden and may begin to pop. Remove the wheat berries and place in a pressure cooker.

Add the brown rice and water. Mix and place the uncovered pressure cooker over a low flame until the water just begins to boil. Add the sea salt, cover the cooker, and bring up to pressure on a high flame. When the pressure is up, place a flame deflector under the cooker and reduce the flame to medium-low. Cook for 45 to 50 minutes. Remove the cooker from the flame and allow the pressure to come down. Remove the cover and allow the rice and wheat to sit for 4 to 5 minutes. Remove the grains and place in a serving bowl.

Brown Rice with Whole Rye

Whole rye berries can be soaked and cooked with brown rice for a dish with a delightfully chewy texture.

3 1/2 cups organic brown rice, washed
1/2 cup rye, washed and soaked for 6 to 8 hours
4 1/2 cups water, including water used to soak rye
small pinch of sea salt

Place the brown rice, rye, and water in a pressure cooker and mix. Place the uncovered pressure cooker over a low flame until the water just begins to boil. Add the sea salt, cover the cooker, and turn the flame to high. When the pressure is up, reduce the flame to medium-low and place a flame deflector under the cooker. Allow to cook for 45 to 50 minutes. Remove the cooker from the flame and allow the pressure to come down. Remove the cover and allow the rice and rye to sit for 4 to 5 minutes. Remove and place in a serving bowl.

Brown Rice with Whole Oats

Whole oats can either be soaked or dry roasted prior to cooking with brown rice.

> 3 1/2 cups organic brown rice, washed
> 1/2 cup whole oats, washed and soaked for 6 to 8 hours
> 4 1/2 cups water, including water used to soak oats
> small pinch of sea salt

Place the brown rice, whole oats, and water in a pressure cooker. Mix and place the uncovered cooker over a low flame until the water begins to boil. Add the sea salt and cover the cooker. Place over a high flame and allow to come to pressure. Reduce the flame to medium-low and place a flame deflector under the cooker. Allow to cook for approximately 45 to 50 minutes. Remove the cooker from the flame and allow the pressure to come down. Remove the cover and allow the rice and oats to sit for 4 to 5 minutes before placing in a serving bowl.

Brown Rice with White Rice

> 2 cups organic brown rice, washed
> 1 cup organic white rice, washed
> 4 cups water
> small pinch of sea salt

Place the rice and water in a pressure cooker without the salt or the lid. Place on a low flame until the water just begins to boil. Add the sea salt, cover, and turn the flame up to high. When the pressure comes up, reduce the flame to medium-low. Place a flame deflector under the cooker and cook for 45 to 50 minutes. Remove from the flame and allow the pressure to come down. Remove the lid and let the rice sit for 4 to 5 minutes. Remove the rice and place in a serving bowl.

Brown Rice with Fresh Sweet Corn

Because sweet corn is soft, and not dry and hard like other grains, you do not need to add extra water when cooking it with rice. It adds a delicious sweet flavor to your rice.

3 cups organic brown rice, washed
1 cup fresh sweet corn, removed from the cob
4 1/2 cups water
small pinch of salt

Place the brown rice, sweet corn, and water in a pressure cooker. Add the water and place the uncovered cooker over a low flame just until the water begins to boil. Add the sea salt and place the cover on the cooker. Raise the flame to high and bring up to pressure. When the pressure is up, reduce the flame to medium-low and place a flame deflector under the cooker. Cook for 45 to 50 minutes. Remove the cooker from the flame and allow the pressure to come down. When the pressure is down, remove the cover and allow the rice and corn to sit for 4 to 5 minutes before placing in a serving bowl.

Brown Rice with Sweet Rice

Sweet brown rice is more glutinous than regular brown rice. It adds extra protein, fat, and sweetness to your rice dishes.

2 cups organic brown rice, washed
1 cup organic sweet brown rice, washed (for a softer texture, soak for 6 to 8 hours)
4 1/2 cups water, including water used to soak grains
small pinch of sea salt

Place the brown rice, sweet brown rice, and water in a pressure cooker and mix. Place the uncovered cooker over a low flame just until the water begins to boil. Add the sea salt, cover, and turn the flame to high. When the pressure is up, reduce the flame to medium-low and place a flame deflector under the cooker. Cook for 45 to 50 minutes. Remove from the flame and allow the pressure to come down. Remove the cover and allow the rice and sweet rice to sit for 4 to 5 minutes before placing in a serving bowl.

Long Grain Rice with Millet

Millet was valued in Oriental medicine for its healing properties, especially its beneficial effect on the pancreas. It can be cooked with short, medium, or long grain rice for a variety of flavors and textures.

2 1/2 cups organic long grain brown rice, washed
1/2 cup organic millet, washed
6 cups water
small pinch of sea salt

Place the brown rice, millet, and water in a heavy pot without a cover. Place on a low flame until the water just begins to boil. Add the sea salt, cover, and turn the flame to high. Reduce the flame to medium-low when the water is at a full boil. Place a flame deflector under the pot. Cook for approximately 1 hour. Remove from the flame and place the brown rice and millet in a serving bowl.

Long Grain Rice with Buckwheat

Buckwheat has a strong contractive quality and warming energy. It is delicious when cooked with long grain rice.

2 1/2 cups long grain brown rice, washed
1/2 cup buckwheat groats, washed
6 1/2 cups water
small pinch of sea salt

Place the brown rice, buckwheat, water, and sea salt in a heavy pot. Cover and bring to a boil over a high flame. Reduce the flame to medium-low, place a flame deflector under the pot, and simmer for approximately 1 hour. Remove from the flame and place in a serving bowl.

Brown Rice with Amaranth

Amaranth, a traditional grain from Central America, can be cooked with brown rice for a distinctive taste.

2 1/2 cups organic brown rice, washed
1/2 cup amaranth, washed
4 1/2 cups water
small pinch of sea salt

Place the brown rice, amaranth, and water in an uncovered pressure cooker. Place over a low flame until the water just begins to boil.

Add the sea salt, cover and turn the flame to high. When the pressure is up, reduce the flame to medium-low and place a flame deflector under the cooker. Cook for approximately 45 to 50 minutes. Remove from the flame and allow the pressure to come down. Remove the cover and allow the grains to sit for 4 to 5 minutes before serving.

Brown Rice with Wild Rice

Wild rice gives brown rice dishes a delightfully rich flavor. It is especially popular during holidays.

> **2 cups organic brown rice, washed**
> **1 cup organic wild rice, washed**
> **4 1/2 cups water**
> **small pinch of sea salt**

Place the brown rice, wild rice, and water in an uncovered pressure cooker. Place over a low flame until the water just begins to boil. Add the sea salt, cover the cooker, and turn the flame up to high. When the pressure is up, reduce the flame to medium-low and place a flame deflector under the cooker. Cook for approximately 45 to 50 minutes. Remove from the flame and allow the pressure to come down. Remove the cover and allow the rice and wild rice to sit for 4 to 5 minutes before placing in a serving bowl.

Brown Rice with Quinoa

Quinoa was traditionally used in the Andes. It is high in protein and adds extra energy to brown rice dishes.

> **2 1/2 cups organic brown rice, washed**
> **1/2 cups organic quinoa, washed**
> **4 1/2 cups water**
> **small pinch of sea salt**

Place the brown rice, quinoa, water, and sea salt in a pressure cooker. Cover, place on a high flame, and bring up to pressure. Reduce the flame to medium-low and place a flame deflector under the cooker. Cook for approximately 45 to 50 minutes. Remove from the flame and allow the rice and quinoa to sit for 4 to 5 minutes before placing in a serving bowl.

Brown Rice with Sweet Rice and Millet

2 cups organic brown rice, washed
1/2 cup organic sweet brown rice, washed
1/2 cup organic millet, washed
4 1/2 cups water
small pinch of sea salt

Place the brown rice, sweet brown rice, and millet in a pressure cooker and mix thoroughly. Add the water and place the uncovered pressure cooker over a low flame just until the water begins to boil. Add the sea salt, place the cover on the cooker, and turn the flame to high. When the pressure is up, place a flame deflector under the cooker and reduce the flame to medium-low. Simmer for 45 to 50 minutes. Remove the cooker from the flame and allow the pressure to come down. Remove the cover when the pressure is down and place the cooked grain in a serving bowl.

Brown Rice with Whole Oats and Millet

2 cups organic brown rice, washed
1/2 cup organic whole oats, washed and soaked for 6 to 8 hours
1/2 cup organic millet, washed
4 1/2 cups water, including water used to soak oats
small pinch of sea salt

Place the brown rice, soaked whole oats, and millet in a pressure cooker and mix thoroughly. Add the water used for soaking the oats and fresh water. Place the uncovered pressure cooker over a low flame just until the water begins to boil. Add the sea salt, place the cover on the cooker, and turn the flame up to high. When the pressure is up, reduce the flame to medium-low and place a flame deflector under the cooker. Cook for 45 to 50 minutes. Remove the cooker from the flame and allow the pressure to come down. When the pressure is down, remove the cover. Allow the grain to sit for 4 to 5 minutes before placing in a serving bowl.

Brown Rice with Pearl Barley and Sweet Corn

> 2 cups organic brown rice, washed
> 1 cup pearl barley (hato mugi), washed
> 1 cup sweet corn, removed from the cob
> 4 1/2 cups water
> small pinch of sea salt

Place the brown rice, hato mugi, and fresh sweet corn in a pressure cooker. Add the water and place the uncovered cooker over a low flame until the water just begins to boil. Add the sea salt, cover the cooker, and turn the flame up high. When the pressure is up, reduce the flame to medium-low and place a flame deflector under the cooker. Cook for 45 to 50 minutes. Remove from the flame and allow the pressure to come down. Remove the cover and allow the grain to sit for 4 to 5 minutes before placing in a serving bowl.

Brown Rice with Whole Wheat and Barley

> 1 1/2 cups organic brown rice, washed
> 1/2 cup organic whole wheat berries, washed and soaked 6 to 8 hours
> 1 cup organic barley, washed and soaked 6 to 8 hours
> 4 1/2 cups water, including water used to soak grains
> small pinch of sea salt

Place the brown rice, wheat berries, and barley in a pressure cooker. Add the water and mix thoroughly. Cover the cooker with a bamboo mat, set aside, and allow to soak 6 to 8 hours or overnight. Remove the bamboo mat. Place the sea salt in the cooker and place the cover on the pressure cooker. Place the pressure cooker over a high flame until the pressure is fully up. Reduce the flame to medium-low and place a flame deflector under the cooker. Cook for 45 to 50 minutes. Remove from the flame and allow the pressure to come down. Remove the cover and allow the grain to sit for 4 to 5 minutes before placing in a serving bowl.

3
Brown Rice with Beans

Cooking brown rice with beans creates a rich, satisfying, and nutritionally complete dish. Because beans usually require longer to cook than grains, they often need advance preparation. They are usually soaked for several hours, roasted in a dry-skillet, or par-boiled for several minutes prior to combining them with brown rice. All beans may be soaked prior to cooking, which makes them softer and easier to digest.

To soak beans, first wash them in cold water, and then place them in a bowl and add enough cold water to cover. Let them soak for 6 to 8 hours. Remove and drain. If you are cooking azuki beans, black soybeans, or chickpeas, you can use the water used for soaking as part of the water measurement. The water used for soaking other beans may be discarded.

Some beans, such as black or yellow soybeans, produce foam when cooked. If you roast them first in a dry skillet, foam will not appear and the beans stay firmer during cooking. This method produces a deliciously sweet dish. If you par-boil the beans for 20 minutes prior to combining them with brown rice, and use the cooking water as part of the final water measurement, this produces a brightly colored dish.

Although the recipes in this chapter call for the use of a pressure cooker, they can just as easily be prepared by boiling in a pot with a tight-fitting lid. Boiling is recommended for those seeking lighter, more relaxing energy in their cooking.

Brown Rice with Azuki Beans

2 cups organic brown rice, washed
1 cup organic azuki beans, washed and soaked 6 to 8 hours
4 1/2 cups water, including water used to soak azuki beans

small pinch of sea salt

Drain the water from the soaked azuki beans and set aside. Place the beans and brown rice in a pressure cooker. Add the water used to soak the beans plus fresh water, according to the amount suggested above. Mix the brown rice and beans. Place the uncovered pressure cooker over a low flame until the water just begins to boil. Add the sea salt, cover, and turn the flame to high. Reduce the flame to medium-low when the pressure is up. Place a flame deflector under the cooker and cook for 45 to 50 minutes. Remove from the flame, allow the pressure to come down, and remove the cover. Allow the rice and beans to sit for 4 to 5 minutes before placing in a serving bowl.

Brown Rice with Black Soybeans

Black soybeans have a thin and delicate skin and need to be washed in a different manner than other beans to prevent the skins from coming off. Take a clean, damp kitchen towel and place the beans in the middle of it. Fold the towel over the beans so that they are completely covered with the towel. Rub the beans with a back and forth, side to side motion. Pour the beans into a bowl. Rinse the towel under cold water to remove soil and dust, and squeeze it out. Place the beans in the towel again and rub as before. Repeat this process one or two more times to completely clean the beans. They are now ready to dry-roast.

> **2 1/2 cups organic brown rice, washed**
> **1/2 cup organic black soybeans, washed**
> **4 1/2 cups water**
> **small pinch of sea salt**

After washing the beans, place them in a strainer to drain. Heat a stainless steel skillet and add the beans. With a wooden spoon or bamboo rice paddle, roast the beans by moving them back and forth and side to side. Start with a high flame, and when the water from washing evaporates, reduce the flame to medium-low. Continue roasting until the skin of the beans becomes very tight and splits slightly, showing a small white streak or split in the skin. Remove the beans from the flame and place them in the pressure cooker. Add the rice and water measurement. Mix the rice and beans. Place the uncovered cooker over a low flame until the water just begins to boil. Place the sea salt in the cooker and place the lid on the cooker. Turn the flame up to high and bring up to pressure. Reduce the flame to medium-low and place a flame deflector under the cooker. Cook for 45 to 50 minutes. Remove the cooker from the flame and allow the pressure to come down. Remove the cover and let the rice and beans sit for 4 to 5 minutes before placing in a serving bowl.

Brown Rice with Kidney Beans

If you par-boil kidney beans prior to combining them with rice, and use the cooking water from the beans, your dish will have an attractive red color.

    2 1/2 cups organic brown rice, washed
    1/2 cup organic kidney beans, washed
    4 1/2 cups water, including cooking water from the beans
    small pinch of sea salt

Place the beans in a saucepan, add cold water to cover, and cover the pan. Bring to a boil on a high flame. Reduce the flame to medium-low and simmer for 20 minutes. Remove from the flame and place the beans in a strainer. Drain the cooking liquid and set aside. Place the beans and rice in the pressure cooker and mix. Combine the cooking water with fresh cold water so as to obtain the above water measurement. Place the water in the cooker. Place the uncovered cooker over a low flame until the water just begins to boil. Add the sea salt and place the lid on the cooker. Turn the flame to high and bring up to pressure. Reduce the flame to medium-low and place a flame deflector under the cooker. Cook for 45 to 50 minutes. Remove the cooker from the flame and allow the pressure to come down. Remove the cover and allow the rice and beans to sit for 4 to 5 minutes before placing in a serving bowl.

Brown Rice with Pinto Beans

    2 1/2 cups organic brown rice, washed
    1/2 cup organic pinto beans, washed and soaked 6 to 8 hours or
      overnight, discard water used for soaking
    4 1/2 cups water
    small pinch of sea salt

Place the brown rice, soaked beans, and water in a pressure

cooker. Place the uncovered cooker over a low flame until the water just begins to boil. Add the sea salt and place the cover on the cooker. Turn the flame to high and bring up to pressure. Reduce the flame to medium-low, place a flame deflector under the cooker, and cook for 45 to 50 minutes. Remove from the flame and allow the pressure to come down. Remove the cover and allow the rice and beans to sit for 4 to 5 minutes before placing in a serving bowl.

Chickpea Rice

> **2 1/2 cups organic brown rice, washed**
> **1/2 cup organic chickpeas, washed and soaked for 6 to 8 hours**
>   **or overnight, discard water used for soaking**
> **4 1/2 cups water**
> **small pinch of sea salt**

Combine the brown rice and chickpeas in a pressure cooker. Add the water and place the uncovered cooker over a low flame until the water just begins to boil. Add the sea salt, place the lid on the cooker, and turn the flame up to high. When the pressure is up, reduce the flame to medium-low and place a flame deflector under the cooker. Cook for 45 to 50 minutes. Remove from the flame and allow the pressure to come down. Remove the cover and allow the rice and beans to sit for 4 to 5 minutes before placing in a serving bowl.

Brown Rice with Whole Wheat and Chickpeas

Whole wheat berries add a firm, chewy texture to this dish, while chickpeas provide a rich, satisfying flavor. This dish makes great fried rice if you have leftovers on the following day.

> **2 cups organic brown rice, washed**
> **1/2 cup organic whole wheat berries, washed, soaked or dry-**
>   **roasted, reserve water used for soaking**
> **1/2 cup organic chickpeas, washed and soaked 6 to 8 hours or**
>   **overnight, discard water used for soaking**
> **4 1/2 cups water, including water used to soak wheat**
> **small pinch of sea salt**

Combine the rice, wheat berries, and chickpeas in a pressure cooker and add the water. Place the uncovered cooker over a low flame until the water just begins to boil. Add the sea salt and place the lid on the cooker. Turn the flame to high and bring up to pressure. Reduce the flame to medium-low and cook for 45 to 50 minutes. Remove from the flame and allow the pressure to come down. Remove the lid and allow the rice and beans to sit for 4 to 5 minutes before placing in a serving bowl.

Brown Rice with Lentils

Lentils are low in fat and have a very short cooking time. Simply wash the lentils and combine with brown rice. They will be done at the same time as the rice.

> **2 1/2 cups organic brown rice, washed**
> **1/2 cup green or brown lentils, washed**
> **4 1/2 cups water**
> **small pinch of sea salt**

Place the rice, lentils, and water in a pressure cooker and mix thoroughly. Place the uncovered cooker over a low flame until the water begins to boil. Add the sea salt, place the lid on the cooker, and turn the flame to high. Reduce the flame to medium-low and place a flame deflector under the cooker. Cook for 45 to 50 minutes. Remove from the flame and allow the pressure to come down. Remove the lid and allow the rice and lentils to sit for 4 to 5 minutes before placing in a serving bowl.

Brown Rice with Mung Beans

Most people are familiar with mung beans in their sprouted form. The mung beans themselves can be combined with brown rice.

> **2 1/2 cups organic brown rice, washed**
> **1/2 cup organic mung beans, washed and soaked 6 to 8 hours or**
> **overnight, discard water used for soaking**
> **4 1/2 cups water**
> **small pinch of sea salt**

Mix the rice and beans in a pressure cooker. Add the water and place the uncovered pressure cooker over a low flame until the water just begins to boil. Add the sea salt, place the lid on the cooker, and turn the flame up to high. When the pressure comes up, reduce the flame to medium-low and cook for 45 to 50 minutes. Remove the cooker from the flame and allow the pressure to come down. Remove the lid and allow the rice and beans to sit for 4 to 5 minutes before placing in a serving bowl.

Boiled Brown Rice with Black Turtle Beans

2 1/2 cups organic brown rice, washed
1/2 cup black turtle beans, washed and soaked 6 to 8 hours or
  overnight, discard water used for soaking
6 cups water
small pinch of sea salt

Place the rice, beans, and water in a heavy pot. Mix and place the uncovered pot on a low flame until the water begins to boil. Add the sea salt, cover the pot, and reduce the flame to medium-low. Place a flame deflector under the pot and simmer for 60 minutes. Remove from the flame and allow the rice and beans to sit for 4 to 5 minutes before placing in a serving bowl.

Brown Rice with Great Northern Beans

2 1/2 cups organic brown rice, washed
1/2 cup great northern beans, washed and soaked for
  6 to 8 hours or overnight, discard water used for
  soaking
1/2 cup sweet corn, removed from cob
4 1/2 cups water
small pinch of sea salt

Mix the rice, beans, and corn in a pressure cooker. Add the water and place the uncovered cooker over a low flame until the water just begins to boil. Add the sea salt, place the lid on the cooker, and turn the flame up to high. When the pressure is up, reduce the flame to medium-low and place a flame deflector under the cooker. Cook for 45 to 50 minutes. Remove the cooker from the flame and allow the pressure to come down. Remove the lid and allow the rice and beans to sit for 4 to 5 minutes before placing in a serving bowl.

Brown Rice with Black-eyed Peas, Seitan, and Vegetables

2 1/2 cups organic brown rice, washed
1/2 cup organic black-eyed peas, washed and soaked 6 to 8
  hours, discardwater used for soaking
1/4 cup cooked seitan, cubed or diced
2 Tbsp celery, diced
2 Tbsp onion, diced
2 Tbsp carrot, diced
1 Tbsp parsley, minced, for garnish
4 1/2 cups water
small pinch of sea salt

Place all ingredients except the parsley in a pressure cooker. Cover the cooker and place over a high flame. When up to pressure, reduce the flame to medium-low and place a flame deflector under the cooker. Pressure cook for 45 to 50 minutes. Remove from the flame and allow the pressure to come down. When the pressure is down, remove the lid and let the rice and beans sit for 4 to 5 minutes before serving.

Brown Rice with Yellow Soybeans

2 1/2 cups organic brown rice, washed
1/2 cup organic yellow soybeans, washed and drained
4 1/2 cups water
small pinch of sea salt

Heat a dry stainless steel skillet over a high flame. Place the soybeans in the skillet and roast evenly, by constantly stirring. Reduce the flame to medium-low when the water used to wash the beans has almost evaporated. Continue roasting until the skin of the beans slightly splits. Remove the beans and mix with the rice in a pressure cooker. Add the water and place the uncovered pressure cooker over a low flame until the water begins to boil. Add the sea salt, place the lid on the cooker, and turn the flame to high. When the pressure comes up, reduce the flame to medium-low and place a flame deflector under the cooker. Cook for 45 to 50 minutes. Remove the cooker from the flame and allow the pressure to come down. Remove the lid from the cooker and allow the rice and beans to sit for 4 to 5 minutes before placing in a serving bowl.

4

Brown Rice Combination Dishes

Brown rice can be combined with a variety of fresh natural ingredients. Combining rice with nuts or seeds, for example, produces a deliciously rich, high-protein dish. Depending on the type of nut or seed you use, you can create dishes with a sweeter or slightly bitter taste, and a more crunchy texture. Beans or vegetables can also be added.

Although many of the recipes in this chapter call for the use of a pressure cooker, they are just as delicious and nourishing when prepared by boiling in a pot with a tight-fitting lid. Boiling is especially recommended for those seeking light, relaxing energy in their food.

Chestnut Rice

This dish has a delicious flavor and helps satisfy the craving for sweets. Dry-roasting the chestnuts prior to cooking produces a very sweet flavor. It causes the chestnuts to retain a firmer consistency.

> 2 cups organic brown rice, washed
> 1 cup organic dried chestnuts, washed and drained
> 4 1/2 cups water
> small pinch of sea salt

Heat a skillet and place the damp chestnuts in it. Dry-roast by stirring with a wooden spoon or bamboo rice paddle, in a back and forth and side to side motion, until the chestnuts become slightly golden in color and release a sweet, nutty fragrance. Place the roasted chestnuts in the pressure cooker. Add the brown rice and water. Mix the chestnuts and rice. Place the uncovered cooker over a low flame until the water just begins to boil. Add the sea salt, place the lid on the cooker, and turn the flame to high. When the pressure is up, reduce the flame to medium-low and place a flame deflector under the cooker. Cook for 45 to 50 minutes. Remove the cooker from the flame and allow the pressure to come down. Remove the lid and allow the rice and chestnuts to sit for 4 to 5 minutes before placing in a serving bowl.

Sweet Brown Rice with Chestnuts

> 1 1/2 cups organic sweet brown rice, washed
> 1/2 cup organic dried chestnuts, washed and soaked 3 to 4
>   hours
> 3 cups water, including the water used for soaking the chest-
>   nuts
> small pinch of sea salt

Place the sweet rice, soaked chestnuts, and water in an uncovered pressure cooker. Place over a low flame until the water begins to boil. Add the sea salt, cover, and turn the flame up to high. When the pressure is up, place a flame deflector under the cooker and reduce the flame to medium-low. Cook for 45 minutes. Remove from the flame and allow the pressure to come down. Remove the cover and let sit for 4 to 5 minutes before placing in a serving bowl.

Brown Rice with Almonds

This dish is very nice when served during holidays or on special occasions. It has a slightly crunchy texture and sweet flavor.

> 2 1/2 cups organic brown rice, washed
> 1/2 cup organic almonds, washed
> 4 1/2 cups water
> small pinch of sea salt

Place the almonds in a saucepan, cover with cold water, and bring to a boil. Boil for 1 to 2 minutes. Remove from the flame and place the almonds in a strainer to drain. Discard the cooking water. With your thumb and index finger, squeeze the almonds one by one. The skin will come off the almond very easily. Discard the skins. Repeat until all of the skins have been removed.

Combine the par-boiled, skinned almonds with the brown rice in a pressure cooker. Add the water. Place the uncovered pressure cooker over a low flame until the water begins to boil. Add the sea salt, place the lid on the cooker, and turn the flame to high. When the pressure comes up, reduce the flame to medium-low and place a flame deflector under the cooker. Cook for 45 to 50 minutes. Remove the cooker from the flame and allow the pressure to come down. Remove the lid and allow the rice and almonds to sit for 4 to 5 minutes before placing in a serving bowl.

Brown Rice with Roasted Walnuts or Pecans

> 3 cups organic brown rice, washed
> 1/2 cup walnuts or pecans, roasted and chopped
> 4 1/2 cups water
> small pinch of sea salt

Place the rice and water in an uncovered pressure cooker. Place over a low flame until the water begins to boil. Add the sea salt, place the lid on the cooker, and turn the flame to high. When the pressure comes up, reduce the flame to medium-low and place a flame deflector under the cooker. Cook for 45 to 50 minutes. Remove from the flame and allow the pressure to come down. Remove the lid and let sit for 4 to 5 minutes before placing in a wooden serving bowl. After placing the rice in a bowl, mix the roasted and chopped nuts thoroughly with the rice.

Brown Rice with Lotus Seeds

In traditional Oriental medicine, lotus seeds were believed to promote strength and longevity.

> 2 1/2 cups organic brown rice, washed
> 1/2 cup organic lotus seeds, washed and soaked for 1 to 2
>   hours, reserve the water used for soaking
> 4 1/2 cups water, including water used for soaking the lotus
>   seeds
> small pinch of sea salt

Combine the rice and lotus seeds in a pressure cooker. Add water and place the uncovered cooker over a low flame until the water just begins to boil. Add the sea salt, place the lid on the cooker, and turn the flame to high. When the pressure is up, reduce the flame to medium-low, and place a flame deflector under the cooker. Cook for 45 to 50 minutes. Remove from the flame and allow the pressure to come down. Remove the cover and let the rice and lotus seeds sit for 4 to 5 minutes before placing in a serving bowl.

Brown Rice with Sesame Seeds

Either tan or black seeds can be used in this recipe.

> 3 cups organic brown rice, washed
> 1/4 cup sesame seeds, washed
> 4 1/2 cups water
> small pinch of sea salt

Mix the rice and sesame seeds in a pressure cooker. Add the water and place the uncovered cooker over a low flame until the water begins to boil. Add the sea salt, place the lid on the cooker, and turn the flame up to high. When the pressure is up, reduce the flame to medium-low and place a flame deflector under the cooker. Cook for 45 to 50 minutes. Remove the cooker from the flame and allow the pressure to come down. Remove the lid and let the rice and sesame seeds sit for 4 to 5 minutes before placing in a serving bowl.

Brown Rice with Pine Nuts

This delicious combination can be enjoyed as a special treat.

> 2 1/2 cups organic brown rice, washed
> 1/2 cup organic pine nuts, washed
> 6 cups water
> small pinch of sea salt

Place the rice and pine nuts in a heavy pot. Mix and add the water. Place the uncovered pot over a low flame until the water comes to a boil. Add the sea salt, cover, and reduce the flame to medium-low. Place a flame deflector under the pot. Cook for 60 minutes. Remove the cover and place the rice and pine nuts in a serving bowl.

Chestnut Rice with Walnuts

2 1/2 cups organic brown rice, washed
1/2 cup organic dried chestnuts, washed and dry-roasted
1/2 cup organic walnuts, washed and coarsely chopped
4 1/2 cups water
small pinch of sea salt

Mix the brown rice, roasted chestnuts, and chopped walnuts in a pressure cooker and add the water. Place the uncovered cooker over a low flame until the water begins to boil. Add the sea salt, place the lid on the cooker, and turn the flame to high. When the pressure is up, reduce the flame to medium-low and place a flame deflector under the cooker. Cook for 45 to 50 minutes. Remove the cooker from the flame and allow the pressure to come down. Remove the cover and let the rice sit for 4 to 5 minutes before placing in a serving bowl.

Brown Rice with Sunflower Seeds

3 cups organic brown rice, washed
1/2 cup organic sunflower seeds, washed and dry-roasted
4 1/2 cups water
small pinch of sea salt

Place the rice and water in a pressure cooker. Place the uncovered cooker over a low flame until the water begins to boil. Add the sea salt, place the lid on the cooker, and turn the flame to high. When the pressure is up, reduce the flame to medium-low and place a flame deflector under the cooker. Cook for 45 to 50 minutes. Remove from the flame and allow the pressure to come down. Remove the cover and let the rice sit for 4 to 5 minutes before mixing in the roasted sunflower seeds. Remove the rice and seeds and place in a serving bowl.

Brown Rice with Roasted Pumpkin Seeds

3 cups organic brown rice, washed
4 1/4 to 4 1/2 cups water
1/2 cup pumpkin seeds, roasted
small pinch of sea salt

Place the rice and water in a pressure cooker, without the sea salt or the lid. Place over a low flame until the water just begins to boil. Add the sea salt, cover, and turn the flame up to high. When the pressure is up, reduce the flame to medium-low, place a flame deflector under the cooker, and cook for 45 to 50 minutes. Remove from the flame and allow the pressure to come down. Remove the lid and allow the rice to sit for 4 to 5 minutes. Mix the roasted pumpkin seeds in with the rice. Remove the rice and place in a serving bowl.

Baked Brown Rice with Almonds and Vegetables

2 1/2 cups organic brown rice, washed and dry-roasted
1/2 cup organic almonds, washed, blanched, and skins removed
1/4 cup onion, diced
1/4 cup celery, diced
1/4 cup mushroom, diced
6 cups boiling water
small pinch of sea salt
2 Tbsp parsley, minced

Mix the roasted rice, almonds, onions, celery, and mushrooms and place in a baking dish or casserole. Add the sea salt and water. Cover the baking dish with foil or a tight fitting lid. Preheat the oven to 350 degrees F. Place the covered dish in the oven and bake for 1 hour. Remove from the oven. Remove the foil wrap and mix in the minced parsley. Place in a serving dish.

Black Soybean and Chestnut Rice

2 cups organic brown rice, washed
1/2 cup organic black soybeans, washed and dry-roasted
1/2 cup organic dried chestnuts, washed and dry-roasted
4 1/2 cups water
small pinch of sea salt

Mix the rice, beans, and chestnuts in a pressure cooker and add the water. Place the uncovered cooker over a low flame until the water begins to boil. Add the sea salt, place the lid on the cooker, and turn the flame to high. When the pressure is up, reduce the flame to medium-low and place a flame deflector under the cooker. Cook for 45 to 50 minutes. Remove the cooker from the flame and allow the pressure to come down. Remove the lid and let the rice sit for 4 to 5 minutes before placing in a serving bowl.

Brown Rice with Umeboshi

Umeboshi, or pickled salt plums, give your rice a slightly salty, sour flavor. Since umeboshi are pickled in sea salt, it is not necessary to add salt to this dish.

> 3 cups organic brown rice, washed
> 1 small umeboshi plum
> 4 1/2 cups water

Place the brown rice, umeboshi, and water in a heavy pot, cover, and bring to a boil. Reduce the flame to medium-low, place a flame deflector under the pot and simmer for 1 hour. Remove from the flame. Remove the cover and place the rice in a serving dish.

Brown Rice with Bancha Tea

Brown rice can occasionally be cooked with bancha tea and a small amount of tamari soy sauce (shoyu) for a slightly stronger flavor.

> 3 cups organic brown rice, washed
> 4 1/2 cups mild bancha twig tea, strained and twigs removed
> 1 tsp organic shoyu (soy sauce)
> 1/4 cup chopped parsley, scallion, or chive, for garnish

Place the rice in a pressure cooker and add the bancha tea and tamari soy sauce. Cover the cooker and place over a high flame. Let the pressure come up. Reduce the flame to medium-low and place a flame deflector under the cooker. Cook for 45 to 50 minutes. Remove from the flame and allow the pressure to come down. Remove the lid and let the rice sit for 4 to 5 minutes before mixing in the chopped parsley, scallion, or chives. Place the rice in a serving bowl.

Brown Rice with Dried Shiitake

Dried shiitake mushrooms add a nice light energy to your brown rice dishes.

> 3 cups organic brown rice, washed
> 3 to 4 dried shiitake, soaked for 10 to 15 minutes
> 4 1/2 cups water, including water used for soaking the shiitake
> small pinch of sea salt

After soaking the shiitake mushrooms, remove from the bowl, squeeze out the water, and remove the woody tip of the stem with a knife. Dice the shiitake. Place the brown rice, shiitake, and water in a pressure cooker and mix the shiitake evenly with the rice. Place the uncovered cooker over a low flame until the water begins to boil. Add the sea salt, place the lid on the cooker, and turn the flame to high. When the pressure comes up, reduce the flame to medium-low and place a flame deflector under the cooker. Cook for 45 to 50 minutes. Remove from the flame and allow the pressure to come down. Remove the lid and allow the rice and shiitake to sit for 4 to 5 minutes before placing in a serving bowl.

Shiso Rice

Shiso leaves are available in two different varieties: red and green. They are sometimes referred to as "beefsteak leaves" in English. Fresh green shiso leaves can be finely chopped and mixed with cooked rice. Red shiso leaves are found mostly in pickled form or in the form of a delicious powdered condiment. Red shiso is also found in containers of umeboshi plums. The red leaves are used to give umeboshi their characteristic color. To serve them with rice, simply rinse them under cold water to remove salt and chop. They can then be mixed with cooked brown rice.

> 3 cups organic brown rice, washed
> 4 1/2 cups water
> small pinch of sea salt
> 1/4 cup red, green, or pickled shiso leaves, finely minced

Pressure cook the brown rice as described earlier. When it is done and the pressure comes down, remove the lid of the pressure cooker and mix the minced shiso leaves with the rice. Remove the rice and place in a serving bowl.

Brown Rice with Ginger Pickles

Ginger root is used in cooking in many countries around the world. It is available in natural and macrobiotic food stores in pickled form. Rinse the ginger pickles, dice or mince very finely, and mix with rice to give your rice a delicious, mildly pungent and salty flavor.

> **3 cups organic brown rice, washed**
> **1/4 cup pickled ginger root, finely minced**
> **2 Tbsp minced parsley**
> **4 1/2 cups water**
> **small pinch of sea salt**

Place the brown rice, water, and sea salt in a pressure cooker and cook as instructed previously. When the rice is done, mix in the minced pickled ginger and parsley. Place in a serving bowl.

Brown Rice with Pickled Daikon

Takuan, or pickled daikon, aids digestion. It is salty and must be rinsed before chopping and combining with cooked brown rice. Sliced takuan can also be soaked for several minutes to remove salt before chopping and mixing with cooked rice. Be sure to use the naturally processed, organic takuan that is sold in natural food stores.

> **3 cups organic brown rice, washed**
> **4 1/2 cups water**
> **small pinch of sea salt**
> **1/3 cup organic takuan pickle, rinsed and finely chopped**
> **parsley sprigs, for garnish**

Place the brown rice, water, and sea salt in a pressure cooker and cook as instructed previously. When the rice is done, mix in the chopped takuan pickle. Place in a serving bowl and garnish with parsley sprigs.

Brown Rice with Fresh Mint

3 cups organic brown rice, washed
4 1/2 cups water
small pinch of sea salt
1/4 cup fresh mint, washed and finely minced

Cook the rice as instructed previously, either in a pressure cook-er or boil in a heavy pot. When the rice is done, mix in the minced mint and place in a serving bowl.

Brown Rice with Squash or Hokkaido Pumpkin

Any kind of hard winter squash, peeled or with the skin left on if the skin is not too tough, may be combined with brown rice to give the dish a delicious, naturally sweet flavor and an attractive orange color. Hokkaido pumpkin, sometimes referred to by the Japanese name *ka-bocha*, is especially delicious because it is very sweet and stays firm during cooking.

3 cups organic brown rice, washed
1 cup organic winter squash or Hokkaido pumpkin, sliced into
   1 inch cubes
4 1/2 cups water
small pinch of sea salt

Place the brown rice and squash or pumpkin cubes in the pres-sure cooker. Add the sea salt and water and mix. Cover the cooker and place over a high flame. When the pressure is up, reduce the flame to medium-low and place a flame deflector under the cooker. Cook for 45 to 50 minutes. Remove from the flame and allow the pressure to come down. Remove the cover and gently mix the rice and squash. Let sit in the cooker for 4 to 5 minutes before placing in a serving bowl.

Brown Rice with Watercress

> 3 cups organic brown rice, washed
> 4 1/2 cups water
> small pinch of sea salt
> 1/2 cup watercress, washed and finely chopped

Cook the rice as explained previously. When the rice is done, mix the chopped watercress in thoroughly. Remove and place in a serving bowl. The heat of the cooked rice will be sufficient to slightly cook the watercress.

Brown Rice with Parsley, Chives, or Scallions

> 3 cups organic brown rice, washed
> 4 1/2 cups water
> small pinch of sea salt
> 1/2 cup finely minced parsley, chives, or scallion

Cook the rice as explained previously. When the rice is done, mix in the minced parsley, chives, or scallion. Remove and place in a serving bowl.

Seitan and Vegetable Gomoku (Mixed Rice)

> 2 cups organic brown rice, washed and dry-roasted
> 1 to 2 square inches of kombu, soaked and diced
> 4 pieces dried tofu, soaked for 10 minutes, diced
> 1 ear of sweet corn, removed from cob
> 1/2 cup carrot, diced
> 1/3 cup seitan, cubed
> 1 stalk celery, diced
> 1/4 cup daikon, diced
> 1/4 cup burdock, diced
> 3 cups water

Place all ingredients in a pressure cooker and mix thoroughly. Add the water, place the lid on the cooker, and place over a high flame. When the pressure is up, reduce the flame to medium-low and place a flame deflector under the cooker. Cook for 40 to 45 minutes.

Remove from the flame and allow the pressure to come down. Remove the cover and let the rice and vegetables sit for 4 to 5 minutes before placing in a serving bowl.

Tempeh and Vegetable Gomoku

> 2 cups organic brown rice, washed
> 1/2 cup tempeh, cubed or diced and deep-fried until golden
> 1/4 cup lotus root (fresh or dried), diced
> 4 to 5 dried shiitake mushrooms, soaked, stems removed, and diced
> 2 Tbsp dried daikon, rinsed, soaked 10 minutes, and chopped
> 2 square inches kombu, soaked and diced
> 1 tsp minced scallion root
> 1/4 cup carrot, diced
> 2 Tbsp scallion, chive, or parsley, minced, for garnish
> 3 cups water, including the water used for soaking the shiitake, dried daikon, and kombu

Place all ingredients in a pressure cooker, add water, and mix thoroughly. Place the lid on the cooker and turn the flame to high. When the pressure is up, reduce the flame to medium-low and place a flame deflector under the cooker. Cook for 40 to 45 minutes. Remove from the flame and allow the pressure to come down. Remove the cover and allow the rice and vegetables to sit for 4 to 5 minutes before mixing in the minced scallion, chives, or parsley garnish. Remove and place in a serving bowl.

Tofu and Vegetable Gomoku

> 2 cups organic brown rice, washed and soaked 6 to 8 hours
> 1 cup firm style tofu, cubed and deep-fried until golden brown
> 1/4 cup fresh sweet corn, removed from the cob
> 1/4 cup fresh green beans, sliced in 1 inch lengths
> 1/4 cup carrot, diced
> 2 Tbsp daikon, diced
> 2 Tbsp burdock, diced
> 2 Tbsp celery, diced
> 2 1/2 to 3 cups water per cup of rice (including the water used for soaking the rice)
> 2 square inches kombu, soaked and diced

Place all ingredients in a pressure cooker, add water, and thoroughly mix. Place the lid on the cooker and turn the flame up to high. When the pressure is up, reduce the flame to medium-low and place a flame deflector under the cooker. Cook for 40 to 45 minutes. Remove from the flame and allow the pressure to come down. Remove the lid and allow the rice and vegetables to sit for 4 to 5 minutes before placing in a serving bowl.

Seafood Gomoku

2 cups organic brown rice, roasted
1/4 cup carrot, diced
2 Tbsp celery, diced
1/4 cup daikon, diced
1/4 cup sweet corn, removed from cob
1 Tbsp minced scallion root
2 Tbsp burdock, diced
1/2 cup fresh baby clams, washed
1/2 cup fresh small shrimp, shelled and veins removed
1/4 cup fresh mussels, washed
2 Tbsp minced parsley or scallion, for garnish
5 1/2 to 6 cups water
2 square inches kombu, soaked and diced

Place all ingredients, except for the clams, shrimp, and mussels, in a heavy pot. Add the water, mix thoroughly, and cover. Place over a high flame and bring to a boil. Reduce the flame to medium-low, place a flame deflector under the cooker, and simmer for 50 minutes. Remove the cover, place the clams, shrimp, and mussels in the pot and mix gently. Cover and cook for another 7 to 10 minutes. Remove from the flame, place in a serving bowl, and garnish with the minced parsley or scallion.

Tempeh Paella

2 cups organic long grain brown rice, washed and dry-roasted
1 cup tempeh, cubed and deep-fried or pan-fried
1/4 cup onion, diced
1/2 cup mushroom, diced
2 cloves garlic, minced (onions may be substituted)
1/4 cup green beans, cut into 1 inch lengths

1/4 cup sweet corn, removed from cob
1 Tbsp extra virgin olive or corn oil
4 cups water
small pinch of sea salt
2 Tbsp minced parsley, chive, or scallion, for garnish

Place the oil in a skillet and heat it. Add the garlic and sauté for 1 minute. Add the mushrooms and sauté for 1 to 2 minutes. Place all the ingredients in a heavy pot, except for the garnish. Add the water, mix, and cover. Place over a high flame and bring to a boil. Reduce the flame to medium-low and place a flame deflector under the cooker. Cook for approximately 50 to 55 minutes until the rice is tender and all the liquid has been absorbed. Remove from the flame and remove the cover. Mix in the minced parsley, chives, or scallion and place in a serving bowl.

Seafood and Vegetable Paella

This paella can be served in the pot you use to cook it in. A paella pan, clay nabé pot, or attractive enameled cast iron pot can be used.

2 cups organic long grain brown rice, washed
2 Tbsp extra virgin olive or corn oil
1/4 cup onion, diced
1/2 cup green beans or green peas
1/2 lb medium shrimp, shelled and tails left on
6 small clams, left in shells and washed
6 small mussels, left in shells and washed
2 to 3 cloves garlic, minced
1/2 cup sweet corn, removed from cob
1/4 tsp sea salt
4 cups water
2 Tbsp minced parsley, for garnish
several lemon wedges, for garnish

Place the oil in a skillet and heat. Add the onions and garlic and sauté for 1 to 2 minutes. Place the sautéed vegetables in a heavy pot. Add the rice, green beans or peas, sweet corn, sea salt, and water. Mix, cover, and bring to a boil. Reduce the flame to medium-low and place a flame deflector under the pot. Cook for about 40 minutes. Place the clams, mussels, and shrimp on top of the rice. Cover and cook for another 20 minutes until the rice is done. Remove the cover, garnish with the minced parsley and lemon wedges, and serve from the pan or pot used for cooking the dish.

Brown Rice with Onions and Black Olives

> 2 cups organic long grain brown rice, washed
> 2 cups onion, diced
> 2 to 3 Tbsp corn oil
> 4 cups water
> 1 cup pitted black olives, sliced
> 1 Tbsp parsley, finely chopped, for garnish

Heat the oil in a skillet and add the diced onion. sauté 2 to 3 minutes. Add the rice and sauté for another minute or two. Add the water, cover, and bring to a boil over a high flame. Reduce the flame to medium-low and simmer for 15 to 20 minutes. Add the olives, cover, and continue cooking for another 35 to 40 minutes until the rice is done and all water has been absorbed. Remove the rice and place in a serving dish, gradually mixing in the chopped parsley with each scoop of rice.

Brown Rice with Fresh Shelled Beans

> 2 cups organic brown rice, washed
> 4 cups water
> 1 cup freshly shelled beans (broad, speckle, etc.)
> 1 tsp fresh dill, finely chopped, for garnish
> 1 Tbsp fresh parsley, finely chopped, for garnish
> small pinch of sea salt

Place the brown rice, beans, and water in a pot. Bring to a boil over a low flame. Add the sea salt, cover, and turn the flame to high. Reduce the flame to medium-low and simmer for 50 to 60 minutes. Mix in the dill and parsley. Remove and place in a serving bowl.

Brown Rice with Deep-fried Tofu and Vegetables

2 cups organic brown rice, washed
1 cup deep-fried tofu, cubed
2 Tbsp bonita (dried fish) flakes
1/2 cup onion, diced
1/4 cup celery, diced
1/4 cup carrot, diced
1/2 cup fresh green peas, boiled until tender
4 cups water
small pinch of sea salt

Place the rice, deep-fried tofu, bonita flakes, onions, celery, carrots, water, and sea salt in a heavy pot. Cover and bring to a boil. Reduce the flame to medium-low and simmer for 50 to 60 minutes. Remove the cover. Mix in the cooked green peas. Remove and place in a serving dish.

5
Rice Balls and Sushi

Rice balls are are easy to make and a great way to use leftover rice. They are wonderful for travel, as brown rice keeps very well when coated with nori and stuffed with umeboshi or pickles. When making rice balls for travel, it is best to use leftover rice rather than warm, fresh rice. Leftover rice will keep longer. Rice balls generally come in four shapes: circles, triangles, spheres, and cylinders. In Japan, rice balls are referred to as *musubi* or *onigiri*. Different types of rice balls are shown in the diagram below.

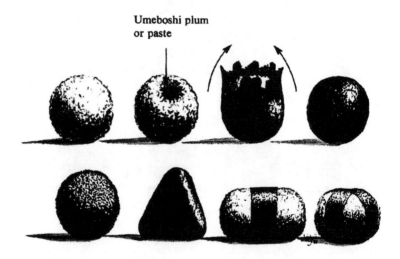

Umeboshi plum
or paste

Rice Balls

> 2 cups (or 2 handfuls) cooked brown rice
> 1 sheet nori
> 1 medium umeboshi plum

Roast the nori with the shiny, smooth side up over a flame. Hold it 10 to 12 inches above the flame, and rotate it until the color changes from black to green (about 3 to 5 seconds). Fold the nori in half and tear along the fold. Then fold in half again and tear so that you have 4 equal-sized pieces of nori (about 3-inches square).

Wet your hands slightly in a dish of water. Take half of the rice in your hands and form into a ball, as if your were making a snowball, or into a triangle by cupping your hands into a V shape. Pack the rice to form a solid ball or triangle. Using your index finger, press a hole into the center of the ball, and place half of the umeboshi plum inside. Then pack the rice again to close the hole. Place 1 square of the toasted nori on the rice ball. Wet your hands slightly and press the nori onto the ball so that it sticks. Take another square of toasted nori and place it on the other, uncovered side. Wet your fingers and press the ball again so that the nori sticks. The rice ball should be completely covered with nori.

Repeat until the rice, nori, and umeboshi are used up.

Sesame Rice Balls

2 cups cooked brown rice
1/4 cup tan or black sesame seeds, roasted
1 medium umeboshi plum

Form the rice into two balls or triangles as instructed above. Poke a hole into the center of the rice ball or triangle with your index finger. Place half of the umeboshi plum in each hole. Press the balls with your hands to close the holes. Roll each ball or triangle in the roasted sesame seeds until completely coated.

Powdered Kombu Rice Bales

2 cups cooked brown rice
1/2 cup tororo kombu (shaved white kombu)

Divide the rice into four equal portions. Shape the rice into cylindrical bales as shown above. Finely chop the tororo kombu. Roll each cylindrical bale in the chopped tororo kombu to completely coat it.

Shiso-leaf Rice Balls

2 cups cooked brown rice
4 shiso leaves (special ones for making rice balls), rinsed

Divide the rice into four equal portions. Mold the rice into circles, triangles, spheres, or cylinders. Take 1 shiso leaf and wrap it around each rice ball and press until it adheres. Repeat until all ingredients are used up.

Pan-fried Azuki Rice Balls

In this recipe, we use leftover brown rice with azuki beans to make delicious deep-fried rice balls.

2 cups cooked brown rice with azuki beans
light or dark sesame oil
organic shoyu (soy sauce)

Form the leftover azuki rice into 4 cylindrical shaped bales. Place a small amount of oil in a skillet and heat. Place the azuki rice bales in the pan and fry until slightly browned on one side. Sprinkle 1 to 2 drops of organic shoyu on the rice bales. Turn the rice bales over and pan-fry the other side until slightly browned. Remove and serve.

Pan-fried Rice Balls

2 cups cooked brown rice
2 Tbsp chive or parsley, finely minced
light or dark sesame oil
small amount of pureéd light or mellow miso, pureéd

Mix the chives or parsley with the cooked rice. Form the rice into small or large triangles. Use your fingers to spread the pureéd miso (sparingly, since miso is salty) on both sides of each triangle. Place a small amount of oil in a skillet and heat. Place the triangles in the skillet and fry until slightly browned. Turn the triangles over and brown the other side. Remove and serve.

Deep-fried Rice Balls

2 cups cooked brown rice
light sesame or safflower oil, for deep-frying
1/4 cup organic shoyu (soy sauce)
1/2 cup water
1 tsp fresh ginger, grated

Form the rice into small balls about the size of a golf ball. Place the oil in a heavy pot suitable for deep-frying. Be sure you have about 2 to 3 inches of oil in the pot. Heat the oil. To test the temperature of the oil, drop a grain of rice in it. If the rice sinks to the bottom and stays there, the oil is not hot enough. When the rice sinks to the bottom and rises to the top almost immediately, the oil is ready to use for deep-frying.

Be careful not to let the oil become too hot, otherwise, it will start to smoke. When the oil has reached the correct temperature, drop several rice balls in it and fry until golden. Remove and place on a paper towel to drain.

To make a dipping sauce, place the organic shoyu, water, and ginger in a saucepan and bring almost to a boil. Turn the flame to very low, and simmer 1 to 2 minutes. Pour the dipping sauce in a small bowl or into individual bowls. The rice balls can be placed in the dipping sauce before eating.

Sushi

Sushi is a traditional Japanese dish that has become very popular lately on an international scale. There are many types of sushi. Sushi can be simply a salad of rice and vegetables that is called *chirashi zushi* or *gomoku zushi*. It can be rice with vegetables, pickles, or fish inside rolled in nori and sliced into rounds.This type of sushi is referred to as *nori maki* or *maki zushi*. There can also be maki without nori or with the nori rolled into a spiral on the inside of the roll. Sushi can also be small bales of rice with raw or cooked slices of fish placed on top.

Sushi is a great picnic food. It is wonderful when traveling, and great for special meals or holidays. It is quick and easy to make and can serve as a snack. The best sushi is made from rice that has cooled to room temperature. Leftover rice can also be used. Do not be discouraged if your first attempt at making sushi is a little frustrating and not so attractive. Practice makes perfect. With a little practice, you will be able to make delicious and appetizing sushi.

> **2 sheets nori, toasted**
> **2 to 3 cups cooked brown rice**
> **1 carrot, cut into 1/4 inch thick lengthwise strips**
> **4 scallions, roots removed**
> **umeboshi paste or umeboshi plums**

Step 1: Roast the rough, dull side of a sheet of nori over a high flame, being careful not to burn it, until it turns from a purple, black color to green. Place the sheet of nori, with the smooth shiny side down, on a bamboo sushi mat. Wet both hands with a little cold water and spread 1 to 1 1/2 cups of leftover cooked rice evenly on the nori. Leave about 1/2 to 1 inch of the bottom part of the nori, the part closest to you, uncovered by rice. Similarly, leave about 2 inches at the top of the sheet, the part farthest from you, uncovered by rice.

Step 2: Slice the carrots into lengthwise strips about 8 to 10 inches long and about 1/4 inch thick. Place the strips in a small amount of boiling water, cover, and boil for 1 to 2 minutes. Remove the carrots and place them on a plate to cool. Place the scallions in the same boiling water, cover, and cook 1 minute. Remove and place on the plate with the carrot strips.

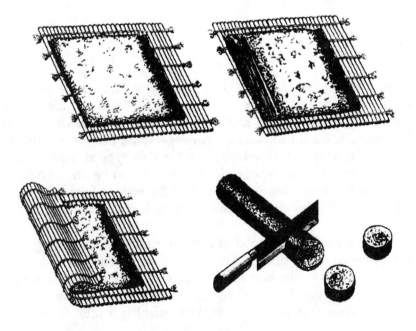

Step 3: Take a small amount of umeboshi paste and spread it evenly across the width of the rice about 1 1/2 to 2 inches from the bottom of the sheet so that it is almost centered on the rice. You can also take small pieces of umeboshi plum and make a line in the same fashion. Use the umeboshi paste moderately, as it is salty. Next, take two carrot strips and lay them on top of the umeboshi paste or plum. Then take two scallions and lay them on top of the carrot strips. You should now have umeboshi, carrots, and scallions lying in a straight line across the width of the nori.

Step 4: Use the bamboo mat to roll up the rice and nori, pressing firmly against the back of the mat with your thumbs, and tucking and rolling forward with your fingers on the nori and vegetables. Roll up until you are about 1 inch from the top of the nori. Wet your fingers again and moisten the end of the nori across the entire width. Continue to roll into a round log or cylindrical shape. The vegetables should be fairly well centered in the roll. Roll the mat completely around the roll and press firmly but gently to seal the nori together. The ends of the rolls are uneven as you can see by looking at either end of the sushi mat. To make the ends even, so that you get eight equal-sized pieces of sushi out of each roll, take a teaspoon of rice and pack it into the ends of the roll. Remove the sushi mat from the roll and set it aside.

Step 5: Wet a very sharp knife and slice the roll in half. Next, slice each half in half. You will now have 4 quarters of a roll. Slice each quarter in half, so that you have 8 equally-sized pieces of sushi.

Step 6: Arrange the sushi rounds (or maki) on a serving platter with the cut end facing up, showing the rice and vegetables.

Step 7: Repeat the above process with the remaining ingredients, so that you have 16 pieces of maki arranged on the platter. Garnish and serve.

Sushi is often served with a dipping sauce made with a small amount of organic shoyu, water, and a little ginger juice or grated daikon. However, a dipping sauce is not really necessary for brown rice (rather than white rice) sushi, or if you are not using fish. Wasabi a hot, green Japanese mustard, is sometimes added to the dip sauce or placed inside the sushi. However, wasabi is not necessary when you are using brown rice and vegetables in your sushi.

Tempeh and Sauerkraut Maki

2 to 3 cups cooked brown rice
2 sheets nori, toasted
1/4 lb tempeh, cut into 1/4 inch thick lengthwise strips
1/4 cup sauerkraut, drained
1/2 tsp natural mustard
4 scallions, roots removed
4 carrot strips, sliced into 1/4 inch thick lengthwise strips
2 tsp tan or black sesame seeds, roasted
water
organic shoyu (soy sauce)
light or dark sesame oil

Place a small amount of oil in a skillet and heat up. Pan-fry the tempeh strips on both sides until golden brown. Add several drops of organic shoyu and add water to almost cover the tempeh. Bring to a boil, cover the skillet, and reduce the flame to medium-low. Simmer for 15 to 20 minutes. Remove the cover, turn the flame to high, and cook off all remaining liquid. Remove the tempeh and place on a plate.

Place a small amount of water in a skillet, cover, and bring to a boil. Remove the cover and place the carrot strips in the skillet. Cover and boil for 1 1/2 minutes. Remove and place on the plate with the tempeh. Place the scallions in the same boiling water, cover, and cook for 1 minute. Remove and place on the plate.

Place the nori on the sushi mat as instructed above. Spread the cooked rice on the nori. Take half of the mustard and draw a straight line across the width of the rice so that it is almost in the center of the sheet. Sprinkle half the roasted sesame seeds on top of the mustard. Take half of the tempeh, half of the carrot strips, and half of the scallions, and lay them in a straight line on top of the mustard and sesame seeds. Take half of the sauerkraut and spread it evenly on top of the tempeh and vegetables.

Roll up the sushi as instructed in the previous recipe and slice into eight equal-sized rounds. Arrange on a platter with the cut end facing up.

Repeat the above process, using all the remaining ingredients.

Tofu and Vegetable Sushi

2 to 3 cups cooked brown rice
2 sheets nori, toasted
1/3 lb firm style tofu, cut into 1/4 inch thick strips, drained
4 carrot strips, cut into 1/4 inch thick lengthwise strips
8 sprigs watercress
several shiso leaves
light sesame or safflower oil, for deep-frying
water
organic shoyu (soy sauce)
brown rice syrup

Place about 2 inches of oil in a heavy pot for deep-frying and heat. When the oil is hot, place the strips of tofu in the oil and deep-fry until golden brown. Remove the strips and drain on paper towels. Place about 1 inch of water in a saucepan and season with a small amount of organic shoyu for a mild salty flavor. Add a small amount of brown rice syrup for a slightly sweet taste. Cover and bring to a boil. Reduce the flame to medium-low and simmer for 10 to 15 minutes. Remove the tofu strips and allow to cool. Squeeze excess liquid from the tofu strips and set on a plate.

Place a small amount of water in a skillet and bring to a boil. Place the carrot strips in the water, cover, and cook for 1 1/2 minutes. Remove and place on the plate with the tofu strips. Place the watercress in the same water and boil for 50 to 60 seconds. Remove and place on the plate.

Place one sheet of toasted nori on a sushi mat. Spread half of the rice on the sheet of nori as instructed previously. Spread half of the carrots, tofu strips, shiso, and watercress evenly in a straight line across the width of the rice so that it is almost centered.

Roll up, slice, and arrange on a platter. Repeat the process again, using all of the remaining ingredients.

Seitan and Vegetable Sushi

2 to 3 cups cooked brown rice
2 sheets nori, toasted
1/2 cup cooked seitan, sliced into strips
1/4 cup sauerkraut, drained
1/2 tsp natural mustard

2 scallions, roots and hard white bases removed

Place the nori on a sushi mat. Spread the rice evenly on the mat as instructed above. Take half of the mustard and spread it evenly across the width of the rice, about 2 inches from the bottom of the sheet. Take half of the sauerkraut and spread it evenly on top of the mustard. Spread half of the seitan strips and one scallion evenly on top of the sauerkraut. Roll up the roll as instructed above and slice into eight pieces. Arrange on a serving platter. Repeat until all of the remaining ingredients have been used.

Cucumber Sushi

2 to 3 cups cooked brown rice
2 sheets nori, toasted
1/4 cup cucumber, sliced into match sticks
1/2 tsp umeboshi paste

Place a sheet of nori on a sushi mat. Spread the rice on the nori as instructed. Take half of the umeboshi paste and spread it evenly, in a straight line, across the width of the rice about 2 inches from the bottom. Place half of the cucumber slices on top of the umeboshi paste. Roll up the roll and cut into eight slices. Arrange on a serving platter. Repeat until all ingredients have been used up.

Natto Sushi

2 to 3 cups cooked brown rice
2 sheets nori, toasted
2 to 3 Tbsp natto (fermented soybeans)
1/2 tsp natural mustard
2 scallions, hard white bases and roots removed
2 to 3 drops organic shoyu (soy sauce)

Mix the natto, mustard, and soy sauce. Chop the scallions very finely and mix with the natto. Place a sheet of nori on a bamboo sushi mat. Spread the rice on the nori as instructed. Take half of the natto mixture and spread it evenly in a straight line across the width of the rice so that it is almost centered. Roll up the rice and natto, slice, and arrange on a serving platter. Repeat until all of the remaining ingredients have been used.

Daikon Pickle Sushi

2 to 3 cups cooked brown rice
2 sheets nori, toasted
2 strips daikon (takuan) pickle, 8 to 10 inches long by
1/4 inch wide

Place one sheet of nori on a sushi mat. Spread the rice evenly on the nori as instructed. Take one strip of the takuan pickle and place it in a straight line across the width of the rice so that it is almost centered. Roll up the rice and takuan, slice, and arrange on a serving platter. Repeat until all ingredients have been used.

6
Rice Salads and Stuffings

Brown rice can be combined with other natural ingredients to make light and delicious salads. It can also be combined with vegetables and other foods to make a delicious stuffings. Below are some basic recipe ideas.

Basic Rice Salad

This dish is light and refreshing, and is wonderful during the summer. Most of the ingredients are cooked for only a short time prior to mixing.

4 cups cooked brown rice
1 cup deep-fried tofu, cut into very thin slices
1/2 cup carrot, cut into thin match sticks
1/4 cup burdock, shaved
5 shiitake mushrooms, soaked 10 to 15 minutes and sliced thin
1/2 cup green string beans, sliced into thin match sticks
1 sheet nori, toasted and cut into thin strips
2 Tbsp tan sesame seeds, toasted
water
organic shoyu (soy sauce)
sesame oil
rice syrup
lemon juice

Place a small amount of water in a saucepan and bring to a boil. Place the carrot match sticks in the water, cover, and simmer 1 minute. Remove and place on a plate. Place the green beans in the water, cover, and cook for 1 to 1 1/2 minutes. Remove and place on the plate with the carrots, keeping them separate. Place the tofu strips in a saucepan with enough water to just cover. Season the water with a little soy sauce for a mild salt taste. Cover and simmer for 10 minutes, then remove and drain. Place the tofu on a plate with the vegetables.

Place the shiitake in a saucepan with enough water to just cover. Season with a little organic shoyu and brown rice syrup for a mild salty-sweet flavor. Simmer for several minutes until all the liquid has evaporated. Remove and place on the plate with the other ingredients. Place a small amount of sesame oil in a skillet and heat. Add the burdock and sauté' for 1 to 2 minutes. Add enough water to half-cover. Cover and simmer for several minutes until tender. Season with a little soy sauce and simmer for another 4 to 5 minutes. Remove and place on the plate with the other ingredients.

Place the fresh cooked rice in a serving bowl. Attractively arrange the vegetables, tofu, and shiitake on top of the rice. Sprinkle the roasted sesame seeds on top. Take 1/2 fresh lemon and squeeze the juice over the vegetable topping. Serve.

Mixed Rice Salad

    1 cup organic brown rice, washed
    1 cup organic white rice, washed
    4 Tbsp brown rice vinegar
    3 Tbsp brown rice syrup
    small pinch of sea salt
    2 1/2 to 2 3/4 cups water
    1/4 cup pickled ginger slices, chopped very fine
    1 sheet nori, toasted
    2 Tbsp tan sesame seeds, toasted
    2 Tbsp bonito flakes (optional)

Place the rice, water, and sea salt in a pressure cooker, cover, and bring up to pressure. Reduce the flame to medium-low and place a flame deflector under the cooker. Cook for 45 minutes. Remove from the flame and allow the pressure to come down. Remove the cover and allow the rice to sit for 4 to 5 minutes. Remove the rice and place in a mixing bowl. In a saucepan, heat the brown rice vinegar and rice syrup. Allow to cool. Mix the vinegar and rice syrup mixture in with the cooked rice.

Add the chopped ginger slices, toasted sesame seeds, bonito flakes, and mix. Tear the nori into small pieces and mix with the rice. Place the mixed rice in a serving bowl.

Stuffed Deep-fried Tofu

2 cups cooked brown or brown and white rice
5 slices firm style tofu, sliced in 1/2 inch thick slices, drained
sesame or safflower oil, for deep-frying
1 cup water
1 strip kombu, 2 inches long
3 Tbsp brown rice syrup
2 Tbsp natural mirin (sweet cooking saké)
organic shoyu (soy sauce)
1/2 cup burdock, sliced into very thin match sticks
1/2 cup carrot, sliced into very thin match sticks
1 1/2 Tbsp black sesame seeds, toasted
sesame oil, for sautéing
3 Tbsp brown rice vinegar

Heat the oil for deep-frying. Place the drained tofu slices in the hot oil and fry until golden brown. Remove and drain on paper towels. Place the tofu in a saucepan with 1 cup water, kombu, brown rice syrup, mirin, and enough organic shoyu for a slightly salty flavor. Cover and bring to a boil. Reduce the flame to medium-low and simmer for about 10 minutes. Remove, drain, and allow to cool. Cut the deep-fried tofu slices in half, forming a triangular shape. Take a knife and insert it into the sliced edge of the tofu to open up the triangle. With a spoon, carefully scrape out all of the tofu inside the triangle, so that only the deep-fried shell of the tofu remains.

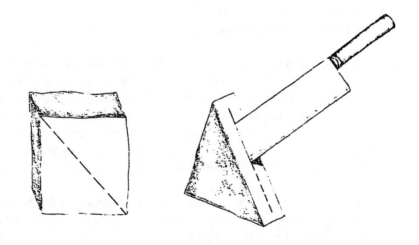

Place a small amount of sesame oil in a skillet and heat. Place the burdock in the skillet and sauté for 2 to 3 minutes. Add enough water to cover the burdock. Lay the carrots on top of the burdock. Cover the skillet and bring to a boil. Reduce the flame to medium-low and simmer for about 7 to 10 minutes until the burdock is tender. Season with several drops of organic shoyu, cover, and cook another 3 to 5 minutes. Remove the cover and turn the flame up. Cook off the remaining liquid.

Place the rice in a mixing bowl. Mix the cooked burdock and carrot, roasted sesame seeds, and brown rice vinegar with the rice. Take a small amount of the rice mixture in your hands and form it into a ball or bale as if making a rice ball. Stuff the rice mixture into one triangular slice of the deep-fried tofu. Place on a serving platter. Repeat until all tofu triangles have been stuffed.

Stuffed Cabbage

4 to 6 cabbage leaves, hard stem at base of the leaf removed
2 cups cooked brown rice
1/2 cup seitan or tempeh, finely diced
1/2 cup onion, diced
1/2 cup carrot, diced
1/4 cup celery, diced
1 Tbsp parsley, minced
1 cup sauerkraut
1/2 cup sauerkraut juice
1/2 cup water
1 strip kombu, 3 inches long
organic shoyu (soy sauce)

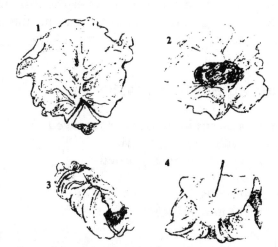

Thoroughly mix the rice, seitan or tempeh, onion, carrot, celery, and parsley in a mixing bowl. Shape the grain and vegetable mixture into oblong rounds or bales. Roll the stuffing up inside the cabbage leaves and fasten with a toothpick to hold the rolls together.

Place the kombu in the bottom of a heavy skillet or shallow pot. Add the sauerkraut juice and water. Place the stuffed cabbage rolls in the skillet. Place the sauerkraut between the rolls. Add a few drops of organic shoyu, cover, and bring to a boil. Reduce the flame to low and simmer for 15 to 20 minutes until soft and tender. Remove the toothpicks from the rolls. Arrange the stuffed cabbage and sauerkraut on a serving platter.

Stuffed Mushrooms

10 large stuffing mushrooms, stems removed
1/2 cup cooked brown rice
1/4 cup seitan, finely minced
1 scallion, finely minced
1/4 cup onion, finely minced
2 Tbsp sauerkraut, finely minced
1/4 cup mochi, grated
1 Tbsp parsley, finely minced
organic shoyu (soy sauce)
sesame oil

Place the rice, seitan, scallion, onion, sauerkraut, and mochi in a mixing bowl. Sprinkle several drops of organic shoyu over the mixture. Mix thoroughly. Stuff each mushroom with about 1 tablespoonful of the mixture. Sprinkle with a little minced parsley. Place the stuffed mushrooms in a baking dish, cover the dish and bake at 350 degrees F. for 15 to 20 minutes until tender and the mochi melts in the stuffing. Remove and place on a serving platter.

Baked Stuffed Acorn Squash

1 acorn squash, cut in half, seeds removed
1/2 cup cooked rice and wild rice
1/2 cup whole wheat bread, cubed
1/4 cup onion, diced
1/4 cup celery, diced
2 Tbsp mushroom, minced

1/4 cup water
organic shoyu (soy sauce)
corn oil, for sautéing

Place a small amount of corn oil in a skillet and heat. sauté the onions for 1 minute. Add the mushroom and celery. Sauté for another 1 to 2 minutes. Sprinkle 4 to 5 drops of organic shoyu over the vegetables. Place the vegetables in a mixing bowl. Add the rice, whole wheat bread, and water. Mix well. Fill each squash half with the stuffing. Place in a baking dish. Cover and bake at 450 degrees F. for about 35 to 40 minutes or until done. Poke with a fork to test. Remove and place on a serving platter.

Stuffed Shiso Leaves

Shiso leaves come pre-packaged in most natural food stores. They can be stuffed and rolled just like grape leaves for a delicious grain dish or snack.

8 to 10 large, fancy shiso leaves, rinsed
1 1/2 to 2 cups cooked brown rice
1/4 cup sunflower seeds, roasted
2 Tbsp chive, scallion, or parsley, finely minced

Place the brown rice, sunflower seeds, and minced chives in a mixing bowl. Mix thoroughly. Take about 1/4 cup of the mixture and form it into oblong rounds like you would if making round-shaped rice bales. Do the same with the remainder of the rice mixture. Wrap or roll the shiso leaves around the stuffing as you would if making stuffed cabbage rolls. Arrange the stuffed shiso leaves on a serving platter.

7
Fried Rice

Fried rice makes a wonderful quick snack. It is a wonderful way to use leftover rice and other foods. The preparation of fried rice can be adjusted to reflect seasonal change. Below are fried rice dishes for each of the seasons, along with ideas for making brown rice burgers and croquettes.

Spring Fried Rice with Wild Vegetables

> 3 cups cooked brown rice
> 2 Tbsp sesame oil
> 1/2 cup chive, finely chopped
> 1/2 cup dandelions, par-boiled 1 minute, finely chopped
> 1/2 cup burdock, shaved or cut into very thin match sticks
> organic shoyu (soy sauce)

Heat the sesame oil in a skillet. Add the burdock and sauté for 1 to 2 minutes. Add the rice, mix, and add several drops of organic shoyu. Cover, reduce the flame to low, and cook until the rice is warm. Stir occasionally to cook evenly. Add the chopped chive and dandelion. Stir, cover, and cook for 1 to 2 minutes. Add a little more organic shoyu for a mild salt flavor. Cover and cook for another minute or so. Mix and place in a serving dish.

Summer Fried Rice with Sweet Corn

> 3 cups cooked brown rice
> 3 ears sweet corn, removed from the cob
> 1/2 cup green beans or peas, sliced thin
> 1/4 cup carrot, diced
> 2 Tbsp sesame seeds, roasted
> 1 sheet nori, toasted
> 2 Tbsp sesame oil
> organic shoyu (soy sauce)

Heat the oil in a skillet. Add the rice and sprinkle several drops of organic shoyu over it. Place the corn, beans, and carrots on top of the rice. Cover the skillet and reduce the flame to low. Cook until the rice is hot. Mix the rice and vegetables. Add several more drops of organic shoyu for a mild taste, mix in the sesame seeds, and cook for another 2 to 3 minutes. Remove and place in a serving dish.

Autumn Fried Rice with Lotus Root

> 3 cups cooked brown rice
> 1 cup fresh lotus root, sliced into thin quarters
> 1/4 cup carrot, sliced into thin quarters
> 1 cup turnip, mustard, or daikon greens
> 2 Tbsp dark (roasted) sesame oil
> 1 tsp ginger juice
> organic shoyu (soy sauce)

Heat the oil in a skillet. Add the lotus root and sauté for 1 to 2 minutes. Place the carrots on top of the lotus root. Place the rice on top of the vegetables. Reduce the flame to low, sprinkle several drops of organic shoyu over the rice, and cover. Cook until the vegetables are done and the rice is hot. Stir occasionally. Place the chopped greens on top of the rice and add a few more drops of organic shoyu. Cover and cook over a medium flame until the greens are tender but still bright green. Sprinkle the ginger juice over the rice and mix well. Remove and place in a serving dish.

Winter Azuki Fried Rice

3 cups cooked brown rice and azuki beans
1 sheet nori, toasted and torn into small pieces
2 Tbsp tan sesame seeds, roasted
1/4 cup onion, diced
1/4 cup squash or pumpkin, diced
1/2 cup scallion or leek, finely chopped
2 Tbsp dark sesame oil
2 to 3 Tbsp water
organic shoyu (soy sauce)

Heat the oil in a cast iron skillet. Add the onions and sauté for 1 to 2 minutes. Place the squash or pumpkin and rice on top of the onion. Add several drops of water and several drops of organic shoyu. Cover and reduce the flame to low. Steam the rice and vegetables until hot. Remove the cover, and place the scallion or leek on top of the rice. Add several more drops of organic shoyu. Cover and cook 1 to 2 minutes until the scallions or leeks are tender and bright green. Remove the cover, mix in the sesame seeds, and place in a serving dish.

All-season Shrimp Fried Rice

3 cups cooked brown and white rice
1/2 lb shrimp, shelled, washed, and chopped
1/4 cup onion, diced
1/4 cup carrot, sliced into match sticks
1/2 cup snow peas, sliced in half
1 Tbsp sesame oil
organic shoyu (soy sauce)

Heat the oil in a skillet. Add the onions and sauté for 1 to 2 minutes. Place the carrots, shrimp, and rice on top of the onions. Sprinkle several drops of organic shoyu over the rice. Cover and reduce the flame to low. Cook until the shrimp and vegetables are done and the rice is hot. Place the snow peas on top of the rice and add several more drops of organic shoyu. Cover and cook another minute or so until the snow peas are tender but still bright green. Mix and place in a serving dish.

Brown Rice and Vegetable Burgers

4 cups cooked brown rice
2 Tbsp parsley, finely minced
1/2 cup onion, finely diced
1/4 cup carrot, finely diced
1/4 cup celery, finely diced
1/4 cup sesame seeds, toasted
organic shoyu (soy sauce)
sesame or corn oil, for frying

Place the rice, parsley, onion, carrot, celery, and sesame seeds in a mixing bowl and mix thoroughly. Form the mixture into 4 to 6 burger-shaped patties.

Oil a skillet or griddle and heat. Place the burgers in the hot skillet. Sprinkle 2 to 3 drops of organic shoyu on top. Fry until golden brown. Turn the burgers over, sprinkle 2 to 3 drops of organic shoyu on the other side, and fry until golden brown. Turn the burgers one more time. Remove and place on a serving platter.

Brown Rice Croquettes with Vegetable-kuzu Sauce

4 cups cooked brown rice
sesame or safflower oil, for deep-frying

Form the cooked rice into 4 to 6 balls or triangles like you would if making rice balls. Heat 2 to 3 inches of oil in a deep-frying pot. Place the rice balls in the hot oil and deep-fry until golden brown. Remove, place on paper towels, and drain the croquettes.

Vegetable-kuzu Sauce

1/2 cup onion, sliced into thick wedges
1/2 cup carrot, sliced into thin diagonals
1 cup broccoli, sliced into flowerettes
1/4 cup celery, sliced into thin diagonals
2 cups water
4 to 5 Tbsp kuzu, diluted in a little cold water

organic shoyu (soy sauce)
ginger juice

Place the water in a pot, cover, and bring to a boil. Add the onions, carrot, and celery. Cover and boil 1 minute. Add the broccoli, cover, and boil 1 minute. Remove the cover. Add the diluted kuzu, stirring constantly to prevent lumping. When the liquid is thick and translucent, reduce the flame to low and season with several drops of organic shoyu for a mild salt flavor. Simmer without a cover for 2 to 3 minutes. Turn the flame off. Squeeze a little fresh ginger juice over the vegetables and mix in. Place 1 to 2 croquettes in each serving bowl. Ladle the vegetable-kuzu sauce over each serving of croquettes and serve.

8
Brown Rice Breakfast Dishes

Brown rice porridges are delicious at breakfast. Breakfast grains are generally cooked with more water than grain dishes served at lunch or dinner. Additional water makes the grains softer and more expanded, and thus more easily digested. A variety of natural foods can be used as seasoning or as condiments when preparing breakfast porridge.

Garnishes also balance the energy in your porridges and other dishes. Choosing the right garnish to balance the flavor, texture, and color of your dishes is an important part of cooking. Once again, recipes that call for pressure cooking can also be boiled.

Soft Rice Porridge

1 cup organic brown rice, washed
5 cups water
small pinch of sea salt

Place the rice, water, and sea salt in a pressure cooker. Cover and place on a high flame. Bring up to pressure. Reduce the flame to medium-low and cook for 45 to 50 minutes. Remove from the flame and allow the pressure to come down. Remove the cover and place in individual serving dishes. Garnish with chopped scallion, chive, or parsley, or sprinkle a little of your favorite condiment over each serving.

Boiled Rice Porridge

1 cup organic brown rice, washed
5 cups water
small pinch of sea salt

Place all ingredients in a heavy pot. Cover and bring to a boil. Reduce the flame to medium-low and simmer for 1 hour. Remove from flame and place in serving bowls. Garnish and serve.

Soft Rice Porridge with Umeboshi

1 cup organic brown rice, washed
1 small to medium umeboshi plum
5 cups water

Place all ingredients in a pressure cooker, cover, and bring up to pressure. Reduce flame to medium-low and simmer for 45 to 50 minutes. Remove from flame and allow the pressure to come down. Remove cover, place in serving bowls, and garnish with toasted nori strips, chopped scallion, and a little gomashio.

Soft Rice Porridge with Pumpkin

1 cup organic brown rice, washed
5 cups water
2 cups Hokkaido pumpkin or buttercup squash, cubed
small pinch of sea salt

Place all ingredients in a pressure cooker, cover, and bring up to pressure. Reduce the flame to medium-low and simmer for 45 to 50 minutes. Remove from flame and allow the pressure to come down. Remove cover and place in serving bowls. Garnish and serve.

Quick Rice Porridge

1 cup leftover brown rice
3 cups water (just to cover rice)

Place rice and water in a pot, cover, and bring to a boil. Reduce the flame to medium-low and simmer for about 20 minutes until soft and creamy. Place in serving bowls and serve with your favorite garnish or condiment.

Ojiya

1 cup organic brown rice, washed
5 cups water
1 inch piece of kombu, soaked and diced
3 to 4 scallion roots, finely minced
3 to 4 scallion tops, thinly sliced
3 to 4 shiitake mushrooms, soaked, stems removed, and sliced,
    add the water used for soaking as part of above water measurement
2 level tsp barley miso, pureéd

Place the rice, kombu, shiitake, scallion roots, and water in a pressure cooker. Cover and bring up to pressure. Reduce the flame to medium-low and cook for 45 to 50 minutes. Remove from flame and allow pressure to come down. Remove cover and place uncovered cooker over a low flame. Add the miso and mix well. Cover cooker with a regular cover (not pressure cooker lid) and simmer, without boiling, for 2 to 3 minutes. Remove cover and place in serving bowls. Garnish with chopped fresh scallion.

Miso Rice Porridge

1 cup organic brown rice, washed
5 cups water
small pinch of sea salt
3 to 4 shiitake mushrooms, soaked, stems removed, and diced,
    add the water used for soaking as part of above liquid measurement
1 cup daikon, quartered and sliced thin
1/4 cup celery, sliced on a thin diagonal
1/2 cup squash, cubed
1/2 cup carrot, diced
1/4 cup cabbage, diced
2 level tsp barley miso, pureéd
chopped scallion, chive, or parsley, for garnish

Place the rice, water, sea salt, and shiitake in a pressure cooker. Cover and bring up to pressure. Reduce flame to medium-low and cook for 45 minutes. Remove from flame and allow pressure to come down. Remove cover. Add the daikon, squash, carrot, and cabbage. Cover with a regular lid, not a pressure cooker lid, and bring to a boil. Reduce the flame to medium-low and simmer several minutes until the vegetables are tender. Reduce the flame to low, add the miso, and mix well. Simmer, without boiling, for 2 to 3 minutes. Place in serving bowls and garnish.

Soft Rice Porridge with Raisins

> 1 cup leftover brown rice
> 3 cups water
> 1/4 cup organic raisins
> 1/4 cup roasted sunflower or pumpkin seeds, for garnish

Place all ingredients in a pot, cover, and bring to a boil. Reduce the flame to medium-low and simmer for 20 minutes until soft and creamy. Place in serving bowls and garnish with toasted sunflower or pumpkin seeds.

Sweet Rice Porridge with Chestnuts

> 1 cup organic sweet brown rice, washed
> 1/2 cup organic dried chestnuts, soaked 3 to 4 hours or dry-
>   roasted and soaked 15 minutes, add the water used for soak-
>   ing as part of above water measurement
> pinch of sea salt
> 5 cups water, including water used to soak chestnuts

Place all ingredients in a pressure cooker, cover, and bring up to pressure. Reduce the flame to medium-low and simmer for 45 to 50 minutes. Remove from flame and allow pressure to come down. Remove cover and place in serving bowls. Garnish and serve.

Pan-Fried Mochi

In Japan sweet brown rice is cooked and pounded with a wooden pestle to make mochi, a sticky, taffy-like rice cake. Mochi is delicious at breakfast or as a quick, anytime snack. It can be found at most natural food stores.

6 pieces of mochi, 3 inches by 2 inches
organic shoyu (soy sauce)
chopped scallion, for garnish

Place the mochi in a heated cast iron or heavy stainless steel skillet. Reduce the flame to low, cover, and brown one side of the mochi. Remove the cover, turn the mochi pieces over, and brown the other side. As the other side is browning, the mochi will puff up slightly. When puffed up and both sides are browned, remove and place on a serving platter. Before eating, sprinkle 1 to 3 drops of organic shoyu over the mochi. Garnish with fresh chopped scallion.

Mochi Waffles with Lemon-walnut Syrup

1 lb mochi
1/2 cup brown rice syrup
2 to 3 Tbsp water
1/4 cup walnuts, roasted and finely chopped
2 to 3 tsp lemon juice, to taste

Slice the mochi into quarters, about 3 1/2 inch by 2 1/2 inch by 1/4 inch thick. Place 1 piece of mochi in each section of a dry (do not oil) waffle iron. Cook until puffed up and slightly crispy but not hard and dry. Repeat until all mochi has been cooked. Place on a serving platter. To prepare the topping, place the rice syrup in a saucepan with the water, roasted walnuts, and a little freshly squeezed lemon juice. Place over a medium flame. When the syrup is hot, pour over each waffle.

Mochi Pancakes with Vegetable Filling

1/4 cup onion, diced
1/4 cup mushroom, sliced thin
1/4 cup carrot, sliced into thin match sticks
1/2 cup cabbage, finely shredded
1/2 cup mung bean sprouts
water
dark sesame oil
1/2 lb plain mochi, coarsely grated
1/2 sheet nori, toasted and cut into thin strips
organic shoyu (soy sauce)

Brush a small amount of oil in a skillet and heat. sauté the onion for 1 to 2 minutes. Add the mushrooms and sauté 1 minute. Add the carrot, cabbage, and mung bean sprouts. Add enough cold water to just coat the bottom of the skillet. Cover, reduce the flame to medium-low, and simmer for 3 to 4 minutes. Remove the cover, sprinkle several drops of organic shoyu over the vegetables, and mix in. Cook 1 minute more. Place the cooked vegetables in a bowl.

On a heated, dry pancake griddle, place 1/4 cup coarsely grated mochi, forming a circle or pancake shape. Spread 1 to 2 tablespoonfuls of the sautéed vegetables on top of the mochi. Sprinkle several strips of nori over the vegetables. Next, quickly sprinkle another 1/4 cup grated mochi on top of the vegetables and nori to create a sandwich effect.

By now the bottom layer of the mochi should have melted and browned slightly. Flip the pancake over and brown the other side. The mochi will melt completely, encasing the vegetable filling. Remove and arrange on a serving platter. Repeat until all ingredients have been used. Serve hot.

Mochi Pancakes with Fruit Filling

1/2 lb plain mochi, coarsely grated
1 apple, cored, peeled, and sliced
1 pear, cored, peeled, and sliced
2 Tbsp raisins
1 to 1 1/2 cups apple juice or water
1 to 1 1/2 Tbsp kuzu, diluted
pinch of sea salt

Place the apples, pears, raisins, juice, and sea salt in a saucepan. Cover and bring to a boil. Reduce the flame to medium-low and simmer 3 to 4 minutes until the fruit is tender. Add the kuzu, stirring constantly, until it becomes thick and translucent. Remove from the flame.

On a heated, dry pancake griddle, place 1/4 cup of grated mochi, forming it into a circle like a pancake. Spread 1 to 2 tablespoonfuls of the stewed fruit evenly on top of the mochi. Next, sprinkle another 1/4 cup of grated mochi on top of the stewed fruit. The bottom layer of mochi should now be slightly browned and melted. Flip the pancake over and cook on the other side until browned and the mochi melts. The fruit topping should now be sandwiched between the two layers of melted mochi. Repeat until all ingredients are used. Serve hot. Other fresh or dried fruits may also be used as filling.

Dietary Guidelines

The Planetary Health Food Pyramid shows how whole grains, including brown rice, are the foundation of a healthy diet. The pyramid and accompanying dietary guidelines are based on human tradition, climatic and environmental considerations (for a temperate climate), nutritional balance, and other factors. They may require personal modification. The guidelines are compatible with a balanced macrobiotic, vegetarian, or vegan diet, many traditional ethnic cuisines, and a modern plant-centered way of eating.

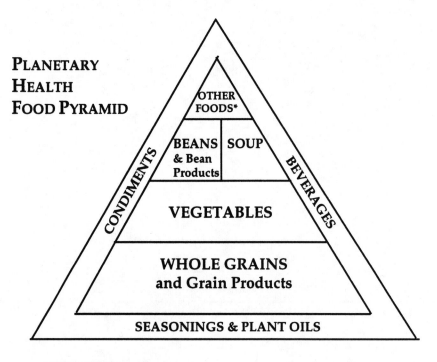

PLANETARY HEALTH FOOD PYRAMID

OTHER FOODS*

BEANS & Bean Products | SOUP

CONDIMENTS

BEVERAGES

VEGETABLES

WHOLE GRAINS and Grain Products

SEASONINGS & PLANT OILS

*Other Foods such as:
Sea Vegetables
Fish and Seafood
Seasonal Fruit
Seeds and Nuts
Natural Snacks
Natural Sweeteners

© 2001 by
Planetary Health, Inc.

1. Whole Grains and Grain Products Approximately half of your daily diet can consist of whole grains, preferably organically grown, non-GE (genetically engineered) varieties. Whole grains include brown rice, barley, millet, whole wheat berries, whole oats, corn, rye, wheat, and buckwheat. Whole grains are the staff of life. Use them as principal food at each meal.

Whenever possible, prepare whole grains daily. Leftovers can be used the following day for breakfast and/or lunch. Whole grain noodles, pasta, Fu (puffed wheat gluten), oatmeal, corn grits, non-yeasted whole sourdough grain bread, and naturally processed grain products are fine for occasional use. Whole grain cookies, crackers, and muffins are best reserved for special occasions, as is seitan (wheat gluten.)

2. Vegetables Vegetables can comprise about a third of your daily intake. Seek fresh, organic (non-GE) produce grown in your region. Vegetables can be cooked in a variety of ways. They can be steamed, boiled, sauteed, stir fried, pickled, pressed, deep fried (tempura), grilled, and served raw.

Vegetables for daily use include roots such as daikon, carrot, turnip, burdock, lotus, parsnip, and radish; round vegetables like winter squash, onion, green cabbage, broccoli, and Brussels sprouts, and leafy greens such as daikon, carrot, and turnip greens, kale, mustard and collard greens, scallion, chive, leek, parsley, and watercress. Shiitake and other mushrooms, green beans, peas, sprouts, summer squash, celery, and lettuce can also be eaten.

3. Beans and Bean Products Beans or bean products can be eaten daily or often in soup, as side dishes, or cooked with rice. Organically grown, non-GE varieties are preferable. Varieties such as azuki, chickpea, and lentil can be used most frequently. Soybean products, such as tofu (fresh and dried), tempeh, and natto (whole fermented soybeans), are also recommended. Artificial, textured vegetable proteins and soy isolates are best avoided.

4. Soup One or two cups or bowls of soup can be eaten daily. Fresh organic vegetables, beans, whole grains, noodles, Shiitake mushrooms, and nori, wakame, and other sea vegetables are delicious in soup. Soups can be lightly seasoned with organic miso (fermented soybean puree), organic shoyu (soy sauce), or white sea salt, and garnished with foods such as sliced scallion, parsley, or chive.

5. Other Foods

• **Sea Vegetables** Edible sea plants can be included regularly. Preferred varieties include nori, wakame, hiziki, arame, kombu, agar, and dulse. Sea vegetables can be used to add flavor and nutrients to soup, vegetable dishes, salads, and condiments.

• **Fish and Seafood** White meat fish is optional but, if desired, can be eaten several times per week. Look for fresh, non-farm raised seafood. Preferred varieties include cod, haddock, halibut, scrod, trout, red snapper, sole, and flounder. Fish can be garnished with foods such as grated daikon, shoyu, ginger, and lemon. Cooking methods include steaming, boiling, and poaching. Broiled and deep-fried fish can be enjoyed from time to time. Red meat fish and shellfish are best reserved for special occasions. Dried fish flakes, or bonito, can be used on occasion to season broth or as garnish.

• **Seasonal Fruit** Cooked, dried, or fresh fruit can be enjoyed when desired. Local fruits, such as apple, pear, peach, apricot, cherry, fresh local berries, cantaloupe, and watermelon are preferable in temperate zones. Juices made from organic local fruits can be enjoyed on occasion. Tropical fruits are best avoided in temperate zones.

• **Seeds and Nuts** Lightly roasted pumpkin, sesame, and sunflower seeds are also fine. A small volume of nuts, such as almond, peanut, walnut, and pecan, can be eaten as snacks. High fat nuts such as Brazil, cashew, and pistachio are best reserved for special occasions.

• **Natural Snacks** Foods such as leftovers, noodles, vegetarian sushi (brown rice wrapped in nori sea vegetable), and mochi (pounded rice taffy) can be enjoyed regularly. Popcorn, puffed whole cereals, brown rice, sea vegetable, and vegetable chips, and rice cakes can also be enjoyed from time to time.

• **Natural Sweeteners** Natural sweeteners such as rice syrup, barley, wheat, and rye malt, amasake (sweet rice milk), and dried chestnuts can be used from time to time in dessert, tea, breakfast cereal, and other dishes.

6. Seasonings and Plant Oils

• **Seasoning** Unrefined white sea salt is recommended for cooking. Traditionally processed miso and shoyu, prepared from organic (non-GE) soybeans, can also be used to season soup and other dishes. Brown rice and umeboshi (pickled plum) vinegar, mirin (rice cooking wine), lemon, and ginger are also fine.

• **Oil** A moderate amount of unrefined sesame oil can be used regularly in sautéing and stir-frying. Other naturally processed vegetable quality oils, such as corn, olive, and sunflower, can also be used.

7. Condiments A variety of traditional natural condiments can be kept on the table and used to add flavor and nutrients to foods. They can be sprinkled on or added to your dishes. They include:

Gomashio (sesame salt made by crushing 20 parts roasted sesame seeds with 1 part roasted sea salt.)

Shiso (perilla) powder

Umeboshi plum

Green nori flakes
Toasted sesame seeds
Tekka
Kinako (soybean flour)
Brown rice vinegar
Umeboshi vinegar
Shoyu (in moderation)
Various combinations of the above

7. Beverages Traditional teas such as bancha, kukicha, barley tea, and brown rice tea can be consumed daily. Organic green tea, Mu tea, corn silk tea, carrot, celery, or vegetable juice, natural amasake and soymilk, and apple juice can also be consumed.

8. Cooking Tips

• Select fresh organic ingredients whenever possible.

• Cook with natural spring or well water.

• Use stainless steel, enamel, ceramic, or cast iron pots, skillets, and utensils. Wooden spoons, chopsticks, cutting boards, and bowls are also recommended. Avoid plastic, aluminum, or chemically coated utensils.

• Wash and put away pots, pans, bowls, plates, and other utensils after using. Keep your kitchen clean and orderly.

• Cook on a gas stove or portable gas burner. If necessary, convert from electricity to gas. Avoid microwave ovens.

• Vary ingredients and cooking methods. Experiment with new dishes and recipes. Learn the art of seasonal cooking. Change ingredients and cooking methods if you travel to a different climate.

• Plan daily menus in advance.

• Cook with a calm, happy, and peaceful mind.

Amberwaves

"In the Far East, the ideogrammic word for 'peace'—wa—*is formed from the ideograms for 'grain' and 'mouth.' Ancient people intuitively knew that a diet based predominantly on grains and vegetables created a peaceful mind and society."*—MICHIO KUSHI AND ALEX JACK, ONE PEACEFUL WORLD

Humanity's true wealth and heritage lie in the diversity of the earth's natural plants and animals; its rich tapestry of cultures, culinary traditions, and customs; and the free flow of ideas, information, and seeds. The Amberwaves Network is being established to bring together farmers; food manufacturers, distributors, retailers, and consumers; health practitioners; environmentalists; religious and spiritual leaders; and individuals, families, and communities to protect and promote whole grains on our planet, focusing on rice as the signal crop to be preserved. The preliminary goals are as follows:

• **Crop Protection:** Declaration of the Sacramento Valley in California and rice-growing regions in Arkansas, Texas, and other areas, as well as restaurants, sushi bars, natural foods stores, supermarkets, churches, schools, hospitals, and other organizations across the country, as GE-Rice Free Zones.

• **Farm Protection:** Implementation of Identity Protection (IP), practical methods and safeguards to separate modified and conventional crops from seed to table; compensation for rice farmers, natural foods distributors, and retail suppliers whose crops or products are contaminated by GE rice in growing, storage, transportation, or processing; and protection for traditional rice strains that have been used for centuries from corporate patenting

• **Health and Environmental Protection:** A moratorium on planting, growing, distributing, or selling GE rice, pending comprehensive tests of its impacts on human health, the environment, organic farming, and the lives of hundreds of millions of farmers and their families who save natural rice seeds for planting each year

• **Consumer Protection:** Mandatory labeling of all GE rice and other foods in the marketplace; product liability to hold GE seed producers, growers, and brokers accountable for their products; and legal redress and health care treatment for consumers who may experience allergenic reactions, toxicity, or other harmful effects from ingesting GE rice

• **Rice Promotion:** Promotion of rice, especially organically grown brown rice, and other whole grains as the foundation of a healthy diet as recently recognized by the new *Dietary Guidelines for Americans*, 2000 edition, that calls upon the American people to consume brown rice and other whole grains daily

What Can I Do?

At a personal and community level, individuals, families, businesses, and other organizations are encouraged to take the following practical steps:

• **Sign the Save Organic Rice Petition and Declaration of a GE-Rice Free Zone:** Sign the petition and declaration (on p. 00) and circulate them among friends, family members, and associates. Copies will be distributed to federal, state, and local officials; the United Nations, World Health Organization, and other international agencies; and to farm groups, the natural foods industry, supermarkets, restaurant chains, and other outlets, showing grassroots opposition.

• **Make a Donation:** Funds are urgently needed to print more literature, start a newsletter, open an office in the Sacramento Valley, and develop a web site. Amberwaves would also like to open offices in Washington, D.C., the United Nations, and send a delegation of organic farmers, environmentalists, and natural foods cooks to visit countries in Asia, Africa, and the Middle East Suggested donation: $10/students and seniors; $25/individuals; $50/families and organizations; $100/supporting donors; and $1000/sustaining donors.

AMBERWAVES
P.O. Box 487
Becket, MA 01223
Tel (413) 623-0012 • fax (413) 623-6042
www.amberwaves.org

Recommended Reading

Books by Wendy Esko

Aveline Kushi's Wonderful World of Salads (Japan Publications, 1989).

The Changing Seasons Cookbook (with Aveline Kushi, Avery Publishing Group, 1985).

Diet for Natural Beauty (with Aveline Kushi, Japan Publications, 1991).

Eat Your Veggies (One Peaceful World Press, 1997).

The Good Morning Macrobiotic Breakfast Book (with Aveline Kushi, Avery Publishing Group, 1991).

Introducing Macrobiotic Cooking (Japan Publications, 1978).

The Macrobiotic Cancer Prevention Cookbook (with Aveline Kushi, Avery Publishing Group, 1988).

Macrobiotic Cooking for Everyone (with Edward Esko, Japan Publications, 1980).

Macrobiotic Family Favorites (with Aveline Kushi, Japan Publications, 1987).

Macrobiotic Pregnancy and Care of the Newborn (with Michio and Aveline Kushi and Edward Esko, Japan Publications, 1984).

The New Pasta Cuisine (with Aveline Kushi, Japan Publications, 1992).

The Quick and Natural Macrobiotic Cookbook (with Aveline Kushi, Contemporary Books, 1989).

Raising Healthy Kids (with Michio and Aveline Kushi and Edward Esko, Avery Publishing Group, 1994).

Soup du Jour (One Peaceful World Press, 1997).

Books by Other Authors

Esko, Edward. *Contemporary Macrobiotics* (1st Books.com, 2000).

Esko, Edward. *Healing Planet Earth* (One Peaceful World Press, 1992).

Esko, Edward. *Notes from the Boundless Frontier* (One Peaceful World Press, 1992).

Esko, Edward. *The Pulse of Life* (One Peaceful World Press, 1994).

Faulkner, Hugh. *Physician Heal Thyself* (One Peaceful World Press, 1992).

Harris-Bonham, Jack. *Medicine Men: A Play about George Ohsawa* (One Peaceful World Press, 1993).

Jack, Alex. *Inspector Ginkgo, The Macrobiotic Detective* (One Peaceful World Press, 1994).

Jack, Alex. *Let Food Be Thy Medicine* (One Peaceful World Press, third edition, 1999).

Jack, Alex. *Out of Thin Air: A Satire on Owls and Ozone, Beef and Biodiversity, Grains and Global Warming* (One Peaceful World Press, 1993).

Jack, Alex. *A Visit to the Land of the Gods* (One Peaceful World Press, 1998).

Jack, Gale and Alex. *Amber Waves of Grain: American Macrobiotic Cooking* (Japan Publications, 1992).

Kushi, Aveline. *Aveline Kushi's Complete Guide to Macrobiotic Cooking* (with Alex Jack, Warner Books, 1985).

Kushi, Michio. *Basic Home Remedies* (One Peaceful World, 1994).

Kushi, Michio. *The Book of Macrobiotics* (with Alex Jack, Japan Publications, revised edition, 1986).

Kushi, Michio. *The Cancer-Prevention Diet* (with Alex Jack, St. Martin's Press, 1983; revised and updated edition, 1993).

Kushi, Michio. *Diet for a Strong Heart* (with Alex Jack, St. Martin's Press, 1985).

Kushi, Michio. *Forgotten Worlds* (with Edward Esko, One Peaceful World Press, 1992).

Kushi, Michio. *The Gospel of Peace: Jesus's Teachings of Eternal Truth* (with Alex Jack, Japan Publications, 1992).

Kushi, Michio. *One Peaceful World* (with Alex Jack, St. Martin's Press, 1986).

About the Author

Wendy Esko, a native of upstate New York and one of the leading natural food cooking teachers in the world, has given thousands of cooking classes over the past twenty-five years. Former director of the Kushi Institute School of Cooking, she has taught macrobiotic cooking in Japan, Italy, and in the Caribbean, and in dozens of cities throughout the United States. Wendy is the author or co-author of nearly twenty books, including *Introducing Macrobiotic Cooking, Macrobiotic Cooking for Everyone, Complete Whole Grain Cookbook, Rice is Nice, Soup du Jour,* and *Eat Your Veggies.* She also offers personal dietary guidance to individuals and families. Mother of eight children and two grandchildren, she lives in Clinton, Michigan and is a marketing assistant with Eden Foods. She is Vice President of Planetary Health/Amberwaves.

Recipe Index

Brown Rice with Pearly Barley and Sweet Corn, 32
Brown Rice with Pickled Daikon, 48
Brown Rice with Pine Nuts, 43
Brown Rice with Pinto Beans, 35
Brown Rice with Quinoa, 29
Brown Rice with Roasted Pumpkin Seeds, 44
Brown Rice with Roasted Walnuts or Pecans, 42
Brown Rice with Sesame Seeds, 43
Brown Rice with Squash or Hokkaido Pumpkin, 49
Brown Rice with Sunflower Seeds, 44
Brown Rice with Sweet Rice and Millet, 31
Brown Rice with Sweet Rice, 28
Brown Rice with Umeboshi, 46
Brown Rice with Watercress, 50
Brown Rice with Wheat Berries, 26
Brown Rice with White Rice, 27
Brown Rice with Whole Oats and Millet, 31
Brown Rice with Whole Oats, 27
Brown Rice with Whole Rye, 26
Brown Rice with Whole Wheat and Barley, 32
Brown Rice with Whole Wheat and Chickpeas, 36
Brown Rice with Wild Rice, 30
Brown Rice with Yellow Soybeans, 39

Chestnut Rice, 40
Chestnut Rice with Walnuts, 44
Chickpea Rice, 36
Cucumber Sushi, 64

Daikon Pickle Sushi, 65
Deep-fried Rice Balls, 59

Long Grain Rice with Buckwheat, 29
Long Grain Rice with Millet, 28

Miso Rice Porridge, 79
Mixed Rice Salad, 67
Mochi Pancakes with Fruit Filling, 82
Mochi Pancakes with Vegetable Filling, 82
Mochi Waffles with Lemon-walnut Syrup, 81

Natto Sushi, 64

Ohsawa Pot Pressure-cooked Brown Rice, 23
Ojiya, 79

Pan-fried Azuki Rice Balls, 58
Pan-fried Mochi, 81
Pan-fried Rice Balls, 58
Powdered Kombu Rice Bales, 57
Pre-soaked Pressure-cooked Brown Rice, 21
Pressure-cooked Roasted Brown Rice, 22

Quick Rice Porridge, 78
Quick-soaked Pressure-cooked Brown Rice, 21

Rice Balls, 56

Seafood and Vegetable Paella, 53
Seafood Gomoku, 52
Seitan and Vegetable Gomoku (Mixed Rice), 50
Seitan and Vegetable Sushi, 62
Sesame Rice Balls, 57
Shiso Rice, 47
Shiso-leaf Rice Balls, 58
Soft Rice Porridge with Pumpkin, 78
Soft Rice Porridge with Raisins, 80
Soft Rice Porridge with Umeboshi, 78
Soft Rice Porridge, 77
Spring Fried Rice with Wild Vegetables, 72
Stuffed Cabbage, 69
Stuffed Deep-fried Tofu, 68
Stuffed Mushroom, 70
Stuffed Shiso Leaves, 73
Summer Fried Rice with Sweet Corn, 73
Sushi, 59
Sweet Brown Rice with Chestnuts, 41
Sweet Rice Porridge with Chestnuts, 80

Tempeh and Sauerkraut Maki, 62
Tempeh and Vegetable Gomoku, 51
Tempeh Paella, 52
Tofu and Vegetable Gomoku, 51
Tofu and Vegetable Sushi, 63

Vegetable-kuzu Sauce, 75

Winter Azuki Fried Rice, 72

Afterword

According to ancient myth, rice originated in Yokota, a village in the deep mountains of Japan. Shortly after the end of World War II, Tomoko Yokoyama, a 28-year-old elementary school teacher from that village came to America. She had no funds, but poor rice farmers scraped together enough money to buy her a third class boat ticket.

With her future husband, Michio, whom she met in New York, Aveline Kushi introduced macrobiotics to America. She started Erewhon, the pioneer natural foods store, and journeyed to California and other regions to persuade farmers to start growing organic brown rice. During her lifetime, she taught thousands of people to cook delicious, healthful meals.

With this tiny woman, the modern organic foods revolution began. In recognition of Aveline's impact on society—as great as any woman, or man, in American history—her pressure cooker is now part of the permanent collection of the Smithsonian Institution in Washington, D.C., along with Ben Franklin's kite, the Wright Brothers airplane, and the Apollo rocket to the moon.

Last year, after returning from a trip to California to meet with rice farmers, I met with Aveline in her home in Brookline, Mass. I explained the threat of genetically engineered (GE) rice. She recalled with sadness recent visits to her village in Japan where chemicals have killed the beautiful insects and tiny fishes that inhabit the rice fields, and she noted some of the farmers even wear gas masks. I explained that natural and organic rice could vanish altogether if new genetic technology spreads. Taking my hand, she said, "Promise me that after I am gone, you will do everything possible to save brown rice. I hope that our teachers and friends will all join together. This is essential for our children's happiness, future generations, and all the little animals that live in the rice fields." Amberwaves was born amid this promise—a promise we can all keep.

Alex Jack
Becket, Massachusetts

Alex Jack is an author, teacher, and president of Amberwaves. Reprinted from "A Promise to Aveline," Amberwaves Journal, *Autumnh, 2001.*